Tucholsky Wagner Zola Scott Sydow Freud Schlegel
Turgenev Wallace Fonatne Schlegel
Twain Walther von der Vogelweide Fouqué Friedrich II. von Preußen
Weber Freiligrath Frey
Fechner Fichte Weiße Rose von Fallersleben Kant Ernst Frommel
Richthofen
Engels Fielding Hölderlin
Fehrs Faber Flaubert Eichendorff Tacitus Dumas
Eliasberg Ebner Eschenbach
Feuerbach Maximilian I. von Habsburg Fock Zweig
Ewald Eliot Vergil
Goethe London
Mendelssohn Balzac Shakespeare Elisabeth von Österreich
Dostojewski Ganghofer
Trackl Lichtenberg Rathenau Doyle Gjellerup
Stevenson Hambruch
Mommsen Tolstoi Lenz Droste-Hülshoff
Thoma Hanrieder
von Arnim Hägele
Dach Verne Hauff Humboldt
Reuter Rousseau Hagen
Karrillon Garschin Hauptmann Gautier
Defoe Baudelaire
Damaschke Hebbel
Descartes
Wolfram von Eschenbach Hegel Kussmaul Herder
Darwin Dickens Schopenhauer
Bronner Melville Grimm Jerome Rilke George
Campe Horváth Aristoteles Bebel Proust
Bismarck Vigny Barlach Voltaire Federer Herodot
Gengenbach Heine
Storm Casanova Tersteegen Grillparzer Georgy
Chamberlain Lessing Langbein Gilm
Brentano Gryphius
Strachwitz Claudius Schiller Lafontaine Kralik Iffland Sokrates
Katharina II. von Rußland Bellamy Schilling
Gerstäcker Raabe Gibbon Tschechow
Löns Hesse Hoffmann Gogol Wilde Vulpius
Luther Heym Hofmannsthal Gleim
Roth Klee Hölty Morgenstern Goedicke
Heyse Klopstock Kleist
Luxemburg Puschkin Homer Mörike
Machiavelli La Roche Horaz Musil
Navarra Aurel Musset Kierkegaard Kraft Kraus
Nestroy Marie de France Lamprecht Kind Kirchhoff Hugo Moltke
Nietzsche Nansen Laotse Ipsen Liebknecht
Marx Ringelnatz
von Ossietzky Lassalle Gorki Klett Leibniz
May vom Stein Lawrence Irving
Petalozzi Knigge
Platon Pückler Michelangelo Kafka
Sachs Poe Liebermann Kock Korolenko
de Sade Praetorius Mistral Zetkin

The publishing house tredition has created the series **TREDITION CLASSICS**. It contains classical literature works from over two thousand years. Most of these titles have been out of print and off the bookstore shelves for decades.

The book series is intended to preserve the cultural legacy and to promote the timeless works of classical literature. As a reader of a **TREDITION CLASSICS** book, the reader supports the mission to save many of the amazing works of world literature from oblivion.

The symbol of **TREDITION CLASSICS** is Johannes Gutenberg (1400 – 1468), the inventor of movable type printing.

With the series, tredition intends to make thousands of international literature classics available in printed format again – worldwide.

All books are available at book retailers worldwide in paperback and in hardcover. For more information please visit: www.tredition.com

tredition was established in 2006 by Sandra Latusseck and Soenke Schulz. Based in Hamburg, Germany, tredition offers publishing solutions to authors and publishing houses, combined with worldwide distribution of printed and digital book content. tredition is uniquely positioned to enable authors and publishing houses to create books on their own terms and without conventional manufacturing risks.

For more information please visit: www.tredition.com

On Laboratory Arts

Richard Threlfall

Imprint

This book is part of the TREDITION CLASSICS series.

Author: Richard Threlfall
Cover design: toepferschumann, Berlin (Germany)

Publisher: tredition GmbH, Hamburg (Germany)
ISBN: 978-3-8491-5452-3

www.tredition.com
www.tredition.de

CHAPTER IV*

ELECTROPLATING AND ALLIED ARTS*

PREFACE

EXPERIMENTAL work in physical science rests ultimately upon the mechanical arts. It is true that in a well-appointed laboratory, where apparatus is collected together in greater or less profusion, the appeal is often very indirect, and to a student carrying out a set experiment with apparatus provided to his hand, the temptation to ignore the mechanical basis of his work is often irresistible.

It often happens that young physicists are to be found whose mathematical attainments are adequate, whose observational powers are perfectly trained, and whose general capacity is unquestioned, but who are quite unable to design or construct the simplest apparatus with due regard to the facility with which it ought to be constructed. That ultimate knowledge of materials and of processes which by long experience becomes intuitive in the mind of a great inventor of course cannot be acquired from books or from any set course of instruction.

There are, however, many steps between absolute ignorance and consummate knowledge of the mechanical arts, and it is the object of the following pages to assist the young physicist in making his first steps towards acquiring a working knowledge of "laboratory arts." However humble the ambition may be, no one can be more keenly alive than the writer to the inadequacy of his attempt; and it is only from a profound sense of the necessity which exists for some beginning to be made, that he has had the courage to air his views on matters about which there are probably hundreds or thousands of people whose knowledge is superior to his own.

Moreover, nothing has been further from the writer's mind than any idea of "instructing" any one; his desire is — if happily it may so befall — to be of assistance, especially to young physicists or inventors who wish to attain definite mechanical ends with the minimum expenditure of time. Most people will agree that one condition essential to success in such an undertaking is brevity, and it is for this reason that alternative methods as a rule have not been given, which, of course, deprives the book of any pretence to being a "treatise." The writer, therefore, is responsible for exercising a certain amount of discretion in the selection he has made, and it is hardly to be hoped that he has in all — or even in the majority of cases — succeeded in recommending absolutely the best method of procedure.

This brings another point into view. Before all things the means indicated must be definite and reliable. It is for this reason that the writer has practically confined himself to matters lying within his own immediate experience, and has never recommended any process (with one or two minor exceptions, which he has noted) which he has not actually and personally carried through to a successful issue. This, although it is a matter which he considers of the highest importance, and which is his only title to a hearing, has unfortunately led to a very personal tone in the book.

With regard to the arts treated of in the following pages, matters about which information is easily acquired — such as carpentering, blacksmithing, turning, and the arts of the watchmaker — have been left on one side. With regard to the last, which is of immense use in the laboratory, there happen to be at least two excellent and handy books, viz. Saunier's Watchmakers' Handbook, Tripplin, London, 1892; and Britton's Watchmakers' Dictionary and Guide.

With regard to carpentering, turning, and blacksmithing, almost any one who so desires can obtain a little practical experience in any village. A short chapter has been devoted to GLASS-BLOWING, in spite of there being an excellent and handy book by Mr. Shenstone (The Methods of GLASS-BLOWING, Rivington) on the subject already in existence. The reason for this exception lies in the fact that the writer's methods differ considerably from those advocated by Mr. Shenstone.

The chapter on opticians' work has had to be compressed to an extent which is undesirable in dealing with so complex and delicate an art, but it is hoped that it will prove a sufficient introduction for laboratory purposes. In this matter the writer is under great obligations to his friend and assistant, Mr. James Cook, F.R.A.S., who gave him his first lessons in lens-making some twenty years ago. To Mr. John A. Brashear of Allegheny, Pa., thanks are due for much miscellaneous information on optical work, which is included verbatim in the text, some of it contained originally in printed papers, and some most kindly communicated to the writer for the purpose of this book. In particular, the writer would thank Mr. Brashear for his generously accorded information as to the production of those "flat" surfaces for which he is so justly famous.

The writer is also indebted to Mr. A. E. Kennelly for some information as to American practice in the use of insulating material for electrical work, and to his friends Mr. J. A. Pollock and Dr. C. J. Martin for many valuable suggestions. For the illustrations thanks are due to Mrs. Threlfall and Mr. James Cook. With regard to matters which have come to the writer's knowledge by his being specifically instructed in them from time to time, due acknowledgment is, it is hoped, made in the text.

With regard to the question as to what matters might be included and what omitted, the general rule has been to include information which the author has obtained with difficulty, and to leave on one side that which he has more easily attained. All the "unities" have been consistently outraged by a deliberate use of the English and metric systems side by side. So long as all the materials for mechanical processes have to be purchased to specifications in inches and feet, it is impossible to use the centimetre consistently without introducing inconvenience. However, everybody ought to, and probably does, use either system with equal facility.

No attempt has been made at showing how work can be done without tools. Though, no doubt, a great deal can be done with inferior appliances where great economy of money and none of time is an object, the writer has long felt very strongly that English physical laboratory practice has gone too far in the direction of starving the workshop, and he does not wish, even indirectly, 'to give any countenance to such a mistaken policy. Physical research is too difficult in itself, and students' time is too valuable, for it to be remunerative to work with insufficient appliances.

In conclusion, the writer would ask his readers to regard the book to some extent as tentative, and as a means to the procuring and organising of information bearing upon laboratory arts. Any information which can be given will be always thankfully received, and the author hereby requests any reader who may happen to learn something of value from the book to communicate any special in-

formation he may possess, so that it may be of use to others should
another edition ever be called for.

CHAPTER I

HINTS ON THE MANIPULATION OF GLASS AND ON GLASS-BLOWING FOR LABORATORY PURPOSES

1. THE art of GLASS-BLOWING has the conspicuous advantage, from the point of view of literary presentation, of being to a great extent incommunicable. As in the case of other delightful arts — such as those treated of in the Badminton Library, for instance — the most that can be done by writing is to indicate suitable methods and to point out precautions which experience has shown to be necessary, and which are not always obvious when the art is first approached. It is not the object of this work to deal with the art of GLASS-BLOWING or any other art after the manner befitting a complete treatise, in which every form of practice is rightly included. On the contrary, it is my wish to avoid the presentation of alternative methods.

I consider that the presentation of alternative methods would, for my present purpose, be a positive disadvantage, for it would swell this book to an outrageous size; and to beginners — I speak from experience — too lavish a treatment acts rather by way of obscuring the points to be aimed at than as a means of enlightenment. The student often does not know which particular bit of advice to follow, and obtains the erroneous idea that great art has to be brought to bear to enable him to accomplish what is, after all, most likely a perfectly simple and straightforward operation.

This being understood, it might perhaps be expected that I should describe nothing but the very best methods for obtaining any proposed result. Such, of course, has been my aim, but it is not likely that I have succeeded in every case, or even in the majority of cases, for I have confined myself to giving such directions as I know from my own personal experience will, if properly carried out, lead to the result claimed. In the few cases in which I have to refer to methods of which I have no personal experience, I have endeavoured to give references (usually taking the form of an acknowledgment), so that an idea of their value may be formed. All methods not particularised may be assumed by the reader to have come within my personal experience.

2. Returning to GLASS-BLOWING, we may note that two forms of GLASS-BLOWING are known in the arts, "Pot" blowing and "Table" blowing. In the former case large quantities of fluid "metal" (technical term for melted glass) are assumed to be available, and as this is seldom the case in the laboratory, and as I have not yet felt the want of such a supply, I shall deal only with "table" blowing. Fortunately there is a convenient book on this subject, by Dr. Shenstone (Rivingtons), so that what I have to say will be as brief as possible, consistent with sufficiency for everyday work. As a matter of fact there is not very much to say, for if ever there was an art in which manual dexterity is of the first and last importance, that art is glass-working.

I do not think that a man can become an accomplished glass-blower from book instructions merely — at all events, not without much unnecessary labour, — but he can learn to do a number of simple things which will make an enormous difference to him both as regards the progress of his work and the state of his pocket.

3. The first thing is to select the glass. In general, it will suffice to purchase tubes and rods; in the case where large pieces (such as the bulbs of Geissler pumps) have to be specially prepared by pot-blowing, the student will have to observe precautions to be mentioned later on. There are three kinds of glass most generally employed in laboratories.

4. Soft Soda Glass,

obtained for the most part from factories in Thuringia, and generally used in assembling chemical apparatus. — This glass is cheap, and easily obtainable from any large firm of apparatus dealers or chemists. It should on no account be purchased from small druggists, for the following reasons:-

(a) It is usually absurdly dear when obtained in this way.

(b) It is generally made up of selections of different age and different composition, and pieces of different composition, even if the difference is slight, will not fuse together and remain together unless joined in a special manner.

(c) It is generally old, and this kind of glass often devitrifies with age, and is then useless for blowpipe work, though it may be bent sufficiently for assembling chemical apparatus. Devitrified glass looks frosty, or, in the earlier stages, appears to be covered by cobwebs, and is easily picked out and rejected.

5. It might be imagined that the devitrification would disappear when the glass is heated to the fusing point; and so it does to a great extent, but for many operations one only requires to soften the glass, and the devitrification often persists up to this temperature. My experience is that denitrified glass is also more likely to crack in the flame than good new glass, though the difference in this respect is not very strongly marked with narrow tubes.

6. Flint Glass. —

Magnificent flint glass is made both in England and France. The English experimenter will probably prefer to use English glass, and, if he is wise, will buy a good deal at a time, since it does not appear to devitrify with age, and uniformity is thereby more likely to be secured. I have obtained uniformly good results with glass made by Messrs. Powell of Whitefriars, but I daresay equally good glass may be obtained elsewhere.

For general purposes flint glass is vastly superior to the soft soda mentioned above. In the first place, it is very much stronger, and also less liable to crack when heated — not alone when it is new, but also, and especially, after it has been partly worked. Apparatus made of flint glass is less liable to crack and break at places of unequal thickness than if made of soda glass. This is not of much importance where small pieces of apparatus only are concerned, because these can generally be fairly annealed; and if the work is well done, the thickness will not be uneven. It is a different matter where large pieces of apparatus, such as connections to Geissler pumps, are concerned, for the glass has often to be worked partly in situ, and can only be imperfectly annealed.

Joints made between specimens of different composition are much more likely to stand than when fashioned in soda glass. Indeed, if it is necessary to join two bits of soda glass of different kinds, it is better to separate them by a short length of flint glass; they are more likely to remain joined to it than to each other. A particular variety of flint glass, known as white enamel, is particularly suitable for this purpose, and, indeed, may be used practically as a cement.

7, It is, however, when the necessity of altering or repairing apparatus complicated by joints arises that the advantage of flint glass is most apparent. A crack anywhere near to a side, or inserted joint, can scarcely ever be repaired in the case of soda glass apparatus, even when the glass is quite thin and the dimensions small.

It should also be mentioned that flint glass has a much more brilliant appearance than soda glass. Of course, there is a considerable difference between different kinds of flint glass as to the melting point, and this may account for the divergency of the statements usually met with as to its fusibility compared with that of soda glass. The kind of flint glass made by Messrs. Powell becomes distinctly soft soon after it is hot enough to be appreciably luminous in a darkened room, and at a white heat is very fluid. This fluidity, though of advantage to the practised worker, is likely to give a beginner some trouble.

8. As against the advantages enumerated, there are some drawbacks. The one which will first strike the student is the tendency of the glass to become reduced in the flame of the blow-pipe. This can be got over by proper adjustment of the flame, as will be explained later on. A more serious drawback in exact work is the following. In making a joint with lead glass it is quite possible to neglect to fuse the glass completely together at every point; in fact, the joint will stand perfectly well even if it be left with a hole at one side, a thing which is quite impossible with soft soda glass, or is at least exceedingly unusual. An accident of this kind is particularly likely to happen if the glass be at all reduced. Hence, if a joint does not crack when cold, the presumption is, in the case of soda glass, that the joint is perfectly made, and will not allow of any leak; but this is not the case with flint glass, for which reason all joints between flint glass tubes require the most minute examination before they are passed. If there are any air bubbles in the glass, especial care must be exercised.

9. Hard or Bohemian, Glass. —

This is, of course, used where high temperatures are to be employed, and also in certain cases where its comparative insolubility in water is of importance. It is very unusual for the investigator to have to make complicated apparatus from this glass. Fused joints may be made between hard glass and flint glass without using enamel, and though they often break in the course of time, still there is no reason against their employment, provided the work be done properly, and they are not required to last too long.

10. On the Choice of Sizes of Glass Tube. —

It will be found that for general purposes tubes about one-quarter inch in inside diameter, and from one-twentieth to one-fortieth of an inch thick, are most in demand. Some very thin soda glass of these dimensions (so-called "cylinder" tubes) will be found very handy for many purposes. For phys-ico-chemical work a good supply of tubing, from one-half to three-quarters of an inch inside diameter, and from one-twentieth to one-eighth inch thick, is very necessary. A few tubes up to three inches diameter, and of various thickness-es, will also be required for special purposes.

Thermometer and "barometer" tubing is occasionally required, the latter, by the way, making particularly bad barometers. The thermometer tubing should be of all sizes of bore, from the finest obtainable up to that which has a bore of about one-sixteenth of an inch. Glass rods varying from about one-twentieth of an inch in diameter up to, say, half an inch will be required, also two or three sticks of white enamel glass for making joints.

To facilitate choice, there is appended a diagram of sizes from the catalogue of a reliable German firm, Messrs. Desaga of Heidelberg, and the experimenter will be able to see at a glance what sizes of glass to order. It is a good plan to stock the largest and smallest size of each material as well as the most useful working sizes.

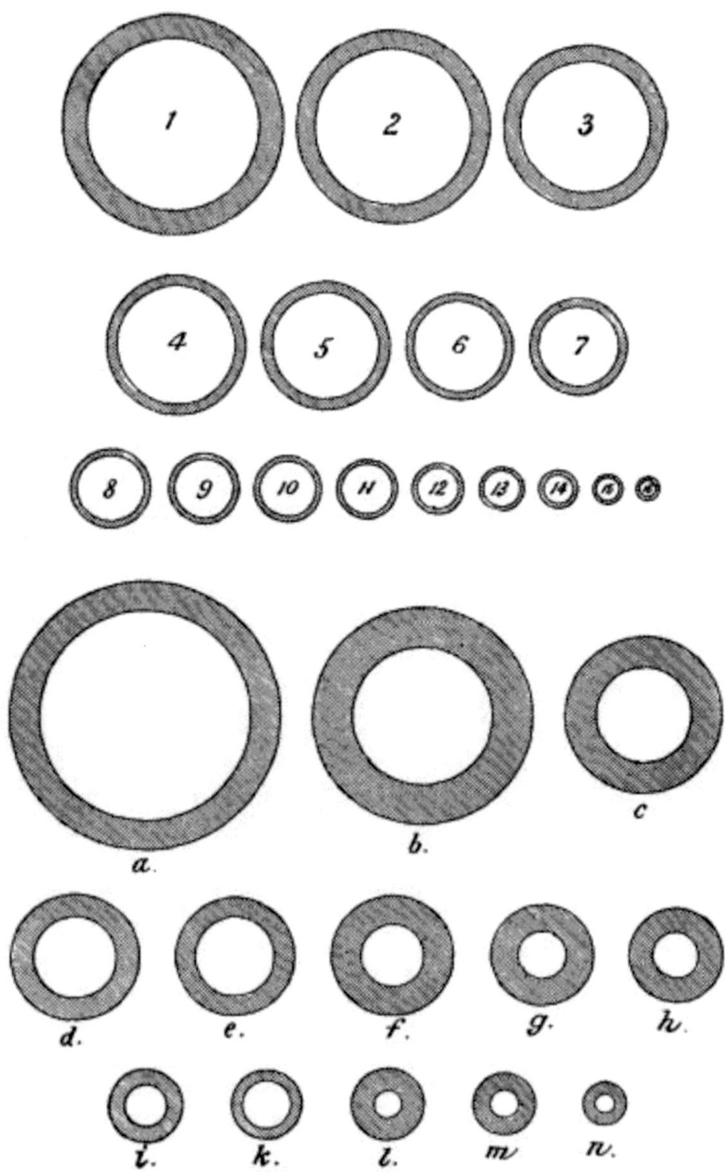

Scanner's note: the scale is not true here. In the original the internal diameter of tube a is about 25mm, and of tube 1 is 20 mm

Fig. 1.

11. Testing Glass. —

"Reject glass which has lumps or knots, is obviously conical, or has long drawn-out bubbles running through the substance." If a scratch be made on the surface of a glass tube, and one end of the scratch be touched by a very fine point of fused glass, say not more than one-sixteenth inch in diameter, the tube, however large it is (within reason), ought to crack in the direction of the scratch. If a big crack forms and does not run straight, but tends to turn longitudinally, it is a sign that the glass is ill annealed, and nothing can be done with it. If such glass be hit upon in the course of blow-pipe work, it is inadvisable to waste time upon it; the best plan is to reject it at once, and save it for some experiment where it will not have to be heated.

The shortest way of selecting glass is to go to a good firm, and let it be understood that if the glass proves to be badly annealed it will be returned. Though it was stated above that the glass should not be distinctly conical, of course allowance must be made for the length of the pieces, and, on the other hand, a few highly conical tubes will be of immense service in special cases, and a small supply of such should be included.

The glass, as it is obtained, should be placed in a rack, and covered by a cloth to reduce the quantity of dust finding its way into the tubes. It has been stated by Professor Ostwald that tubes when reared up on end tend to bend permanently. I have not noticed this with lead glass well supported. Each different supply should be kept by itself and carefully described on a label pasted on to the rack, and tubes from different lots should not be used for critical welds. This remark is more important in the case of soda than of lead glass.

In the case of very fine thermometer tubes it will be advisable to cover the ends with a little melted shellac, or, in special cases, to obtain the tubes sealed from the works. Soda glass can generally be got in rather longer lengths than lead glass; the longer the lengths are the better, for the waste is less.

It is useful to be able to distinguish the different kinds of glass by the colour. This is best observed by looking towards a bright surface along the whole length of the tube and through the glass. Lead glass is yellow, soda glass is green, and hard glass purple in the samples in my laboratory, and I expect this is practically true of most samples. *[Footnote: Some new lead glass I have is also almost purple in hue. If any doubt exists as to the kind of glass, it may be tested at once in the blow-pipe flame, or by a mixture of oils of different refractive indices, as will be explained later.]*

12. The question of the solubility of glass in reagents is one of great importance in accurate work, though it does not always meet with the attention it deserves. It is impossible here to go into the matter with sufficient detail, and the reader is therefore referred to the *Abstracts* of the Chemical Society, particularly for the years 1889 and 1892. The memoir by F. Kohlrausch, *Wied. Ann. xliv.*, should be consulted in the original. The following points may be noted. A method of testing the quality of glass is given by Mylius (*C. S. J. Abstracts*, 1889, p. 549), and it is stated that the resistance of glass to the action of water can generally be much increased by leaving it in contact with cold water for several days, and then heating it to 300 to 400 C. This improvement seems to be due to the formation of a layer of moist silica on the surface, and its subsequent condensation into a resisting layer by the heating. Mylius (*C. S. J. Abstracts,* 1892, p. 411), and Weber, and Sauer (*C. S. J. Abstracts*, 1892, p. 410) have also shown that the best glass for general chemical purposes consists of

Silica, 7 to 8 parts

Lime, 1 part

Alkali, 1.5 to 1.1 parts.

This is practically "Bohemian" tube glass.

The exact results are given in the *Berichte* of the German Chemical Society, vol. xxv. An excellent account of the properties of glass will be found in Grove's edition of Miller's *Elements of Chemistry*.

13. Cleaning Glass Tubes. —

This is one of the most important arts in chemistry. If the tubes are new, they are generally only soiled by dust, and can be cleaned fairly easily — first by pushing a bit of cotton waste through with a cane, or pulling a rag through with string — and then washing with sand and commercial hydro-chloric acid. I have heard of glass becoming scratched by this process, and breaking in consequence when heated, but have never myself experienced this inconvenience. In German laboratories little bits of bibulous paper are sometimes used instead of sand; they soon break into a pulp, and this pulp has a slightly scouring action.

As soon as the visible impurities are removed and the tube when washed looks bright and clean, it may be wiped on the outside and held perpendicularly so as to allow the water film to drain down. If the tube be greasy (and perhaps in other cases) it will be observed that as the film gets thinner the water begins to break away and leave dry spots. For accurate work this grease, or whatever it is, must be removed; and after trying many plans for many years, I have come back to the method I first employed, viz. boiling out with aqua regia.

For this purpose, close one end of the tube by a cork (better than a rubber bung, because cheaper), and half fill the tube with aqua regia; then, having noted the greasy places, proceed to boil the liquid in contact with the glass at these points, and in the case of very obstinate dirt — such as lingers round a fused joint which has been made between undusted tubes — leave the whole affair for twelve hours. If the greasiness is only slight, then simply shaking with hot aqua regia will often remove it, and the aqua regia is conveniently heated in this case by the addition of a little strong sulphuric acid.

The spent aqua regia may be put into a bottle. It is generally quite good enough for the purpose of washing glass vessels with sand, as above explained.

However carefully a tube is cleaned before being subjected to blowpipe operations, it will be fouled wherever there is an opening during the process of heating, unless the extreme tip only of an oxidising flame be employed. Even this should not be trusted too implicitly unless an oxygas or hydrogen flame is employed.

When a tube or piece of apparatus has been cleaned by acid, so that on clamping it vertically, dry spaces do not appear, it may be rinsed with platinum distilled water and left to drain, the dust being, of course, kept out by placing a bit of paper round the top. For accurate work water thus prepared is to be preferred to anything else. When the glass is very clean interference colours will be noticed as the water dries away.

Carefully-purified alcohol may in some cases be employed where it is desired to dry the tube or apparatus quickly. In this case an alcohol wash bottle should be used, and a little alcohol squirted into the top of the tube all round the circumference. The water film drags the alcohol after it, and by waiting a few minutes and then adding a few more drops of alcohol, the water may be practically entirely removed, especially if a bit of filter paper be held against the lower end of the tube. It is customary in some laboratories to use ether for a final rinse, but unless the ether is freshly distilled and very pure, it leaves a distinct organic residue.

When no more liquid can be caused to drain away, the tube may be dried by heating it along its length, beginning at the top (to get the advantage of the reduction of surface tension), and so on all down. It will then be possible to mop up a little more of the rinsing liquid. When the tube is nearly dry a loose plug of cotton wool may be inserted at the bottom. The wool must be put in so that the fibres lie on an even surface inside the tube, and the wool must be blown free from dust. Ordinary cotton wool is useless, from being dusty and the fibres short, and the same remark applies to wadding. Use nothing but what is known as "medicated" cotton wool with a good long fibre.

The tube will usually soon dry of itself when the cover is lifted an inch or so. If water has been used, the air-current may be assisted by means of the water-pump, the air being sucked from the top, so that the wool has an opportunity of acting as a dust filter; a very slow stream of air only must be employed. For connecting the tube to the pump, a bit of India-rubber tube about an inch in diameter, with a bore of about one-eighth of an inch, may be employed. The end of the rubber tube is merely pressed against the edge of the glass.

These remarks apply, with suitable modification, to all kinds of finished apparatus having two openings. For flasks and so on, it is convenient to employ a blowing apparatus, dust being avoided by inserting a permanent plug of cotton wool in one of the leading tubes. The efficiency of this method is greatly increased by using about one foot of thin copper tube, bent into a helix, and heated by means of a Bunsen burner; the hot air (previously filtered) is passed directly into the flask, bottle, or whatever the apparatus may be. This has proved so convenient that a copper coil is now permanently fastened to the wall in one of the rooms of my laboratory.

The above instructions indicate greater refinement than is in general necessary or proper for tubes that have to be afterwards worked by the blow-pipe. In the majority of cases all that is necessary is to remove the dust, and this is preferably done by a wad of cotton waste (which does not leave shreds like cotton wool), followed by a bit of bibulous filter paper. I would especially warn a beginner against neglecting this precaution, for in the process of blowing, the dust undergoes some change at the heated parts of the apparatus, and forms a particularly obstinate kind of dirt.

In special cases the methods I have advocated for removing dirt and drying without covering the damp surfaces with dust are inadequate, but an experimenter who has got to that stage will have nothing to learn from such a work as this.

14. The Blow-pipe. —

I suppose a small book might easily be written on this subject but what I have to say — in accordance with the limitation imposed — will be brief. For working lead glass I never use anything but an oxygas blow-pipe, except for very large work, and should never dream of using anything else. Of course, to a student who requires practice in order to attain dexterity this plan would be a good deal too dear. My advice to such a one is — procure good soda glass, and work it by means of a modification of a gas blow-pipe, to be described directly. The Fletcher's blow-pipes on long stems are generally very inconvenient. The flame should not be more than 5 or

6 inches from the working table at most, especially for a beginner, who needs to rest his arms on the edge of the table to
secure steadiness.

The kind of oxygas blow-pipe I find most convenient is indicated
in the sketch. (Fig. 2) I like to have two nozzles, which will slip on
and off, one with a jet of about 0.035 inch in diameter, the other of
about double this dimension. The oxygen is led into the main tube
of the blow-pipe by another tube of much smaller diameter, concentric with the main tube (Fig. 3, at A). The oxygen is mixed with the
gas during its escape from the inner tube, which is pierced by a
number of fine holes for the purpose, the extreme end being closed
up. The inner tube may run up to within half an inch of the point
where the cap carrying the nozzle joins the larger tube.

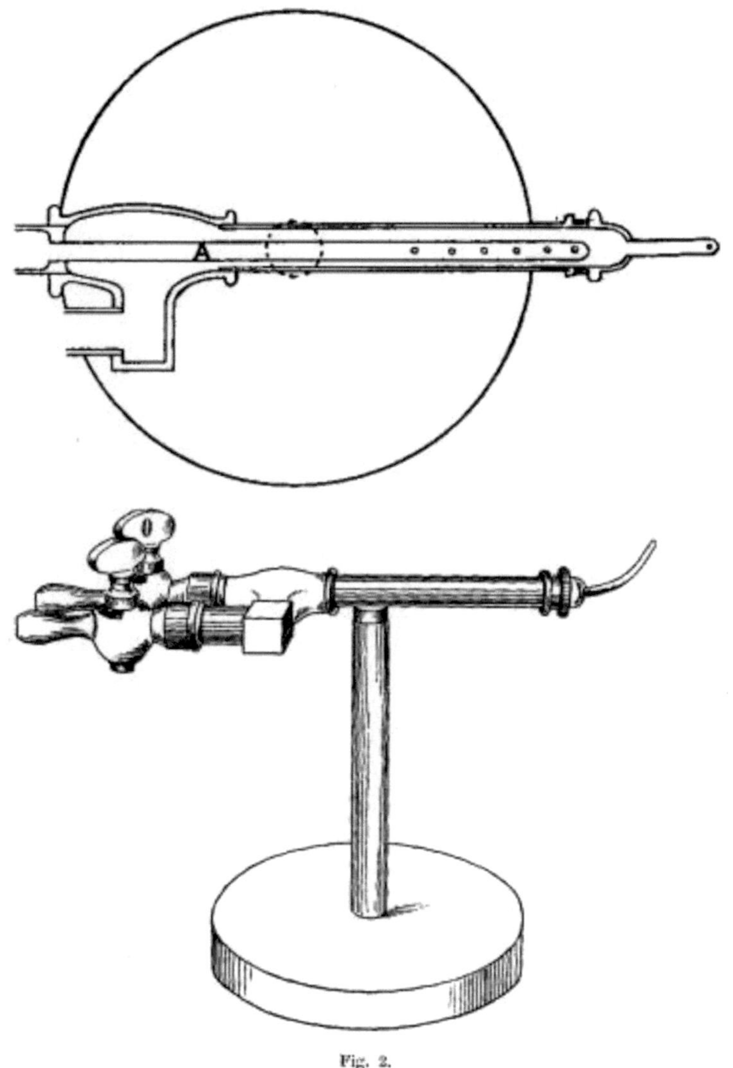

Fig. 2.

Fig. 2.

Fig. 3.

If it is desired to use the blow-pipe for working glass which is already fixed in position to a support, it will be found very advantageous to use a hooked nozzle. The nozzle shown in the sketch is not hooked enough for this work, which requires that the flame be directed 'backwards towards the worker. With a little practice such a flame may be used perfectly well for blowing operations on the table, as well as for getting at the back of fixed tubes.

To warm up the glass, the gas supply is turned full on, and enough oxygen is allowed to pass in to clear the flame. The work is held in front of, but not touching, the flame, until it is sufficiently hot to bear moving into the flame itself. The, work is exposed to this flame until, in the case of lead glass, traces of reduction begin to appear. When this point is reached the oxygen tap is thrown wide open. I generally regulate the pressure on the bags, so that under these circumstances the flame is rather overfed with oxygen. This condition is easily recognised, as follows. The flame shrinks down to a very small compass, and the inner blue cone almost disappears; also flashes of yellow light begin to show themselves — a thing which does not occur when the proportions of the gases are adjusted for maximum heating effect.

For many purposes the small dimensions of the flame render it very convenient, and the high temperature which can be attained at exact spots enables glass to be fused together after a certain amount of mixing, which is an enormous advantage in fusing lead glass on to hard glass. The lead glass should not be heated hot enough to burn, but, short of this, the more fluid it is the better for joints between dissimilar samples.

It will be noticed that the blow-pipe can be rotated about a vertical axis so as to throw the flame in various directions. This is often indispensable.

15. In general the oxygen flame does not require to be delivered under so high a pressure as for the production of a lime light. In England, I presume, most experimenters will obtain their oxygen ready prepared in bottles, and will not have to undergo the annoyance of filling a bag. If, however, a bag is used, and it has some advantages (the valves of bottles being generally stiff), I find that a pressure produced by placing about two hundredweight (conven-

iently divided into four fifty-six pound weights) on bags measuring 3' x 2'6" x 2' (at the thicker end) does very well. To fill such a bag with oxygen, about 700 grms of potassium chlorate is required.

If the experimenter desires to keep his bag in good order, he must purify his oxygen by washing it with a solution of caustic soda, and then passing it through a "tower" of potash or soda in sticks, and, finally, through a calcium chloride tower. This purifying apparatus should be permanently set up on a board, so that it may be carried about by the attendant to wherever it is required. Oxygen thus purified does not seem to injure a good bag — at least during the first six or seven years:

In order to reduce the annoyance of preparing oxygen, the use of the usual thin copper conical bottle should be avoided. The makers of steel gas bottles provide retorts of wrought iron or steel for oxygen-making, and these do very well. They have the incidental advantage of being strong enough to resist the attacks of a servant when a spent charge is being removed.

The form of retort referred to is merely a large tube, closed at one end, and with a screw coupling at the other; the dimensions may be conveniently about 5 inches by 10. The screw threads should be filled with fireclay (as recommended by Faraday) before the joint is screwed up. Before purchasing a bottle the experimenter will do well to remember that unless it is of sufficiently small diameter to go into his largest vice, he will be inconvenienced in screwing the top on and off. Why these affairs are not made with union joints, as they should be, is a question which will perhaps be answered when we learn why cork borers are still generally made of brass, though steel tube has long been available.

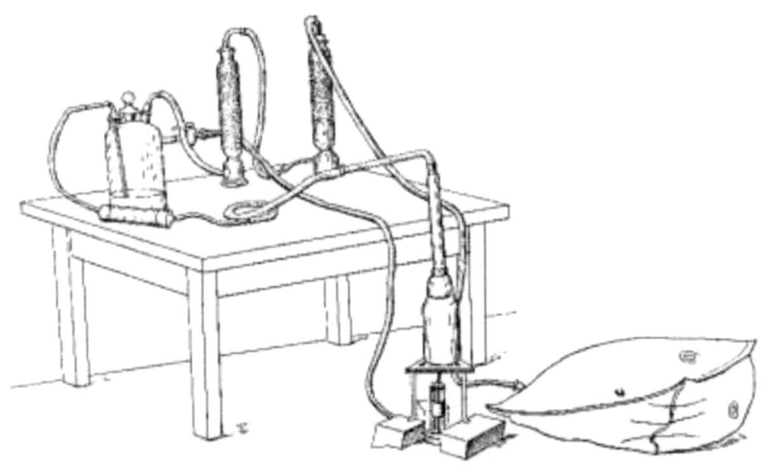

Fig. 4.

These little matters may appear very trivial — and so they are — but the purchaser of apparatus will generally find that unless he looks after details himself, they will not be attended to for him. Whether a union joint is provided or not, let it be seen that the end of the delivery tube is either small enough to fit a large rubber tube connection going to the wash-bottle, or large enough to allow of a cork carrying a bit of glass tube for the same purpose to be inserted. This tube should not be less than half an inch in inside diameter. Never use a new bottle before it has been heated sufficiently to get rid of grease and carbonaceous dirt. A convenient oxygen-making apparatus is shown in Fig. 4, which is drawn from "life."

16. For large blow-pipe work with lead glass I recommend a system of four simple blow-pipes, in accordance with the sketch annexed. I first saw this system in operation in the lamp factory of the Westinghouse Electric Company at Pittsburg in 1889, and since then I have seen it used by an exceedingly clever "trick" glass-worker at a show. After trying both this arrangement and the "brush flame" recommended by Mr. Shenstone, I consider the former the more convenient; however, I daresay that either can be made to work in competent hands, but I shall here describe only my own choice. [Footnote: A brush flame is one which issues from the blow-pipe nozzle shaped like a brush, i.e. it expands on leaving the jet. It is

32

produced by using a cylindrical air jet or a conical jet with a large aperture, say one-eighth of an inch. See Fig. 25.]

As will be seen, the blow-pipe really consists of four simple brass tube blow-pipes about three-eighths of an inch internal diameter and 3 inches long, each with its gas and air tap and appropriate nozzle. Each blowpipe can turn about its support (the gas-entry pipe) to some extent, and this possibility of adjustment is of importance, The air jets are merely bits of very even three-sixteenths inch glass tubing, drawn down to conical points, the jets themselves being about 0.035 inch diameter.

Fig.   5

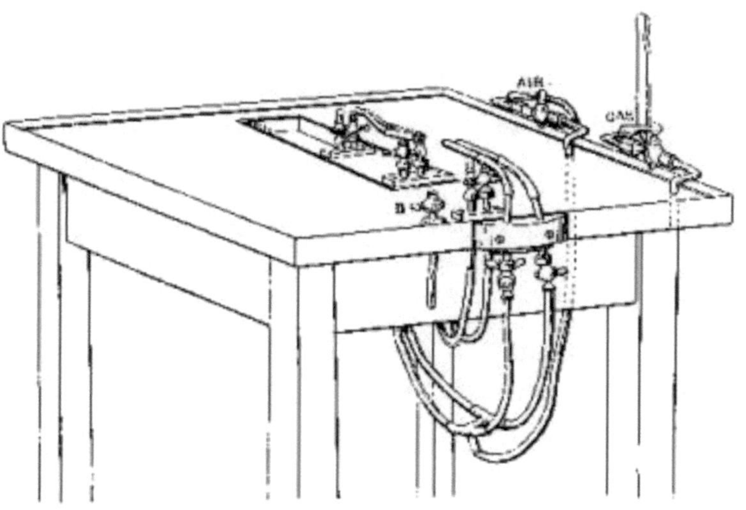

The flames produced are the long narrow blow-pipe flames used in blow-pipe analysis, and arranged so as to consist mostly of oxidising flame. The air-supply does not require to be large, nor the pressure high — 5 to 10 inches of water will do — but it must be very regular. The "trick" glass-blower I referred to employed a foot bellows in connection with a small weighted gasometer, the Westinghouse Company used their ordinary air-blast, and I have generally used a large gas-holder with which I am provided, which is supplied by a Roots blower worked by an engine.

I have also used a "velocity pump" blower, which may be purchased amongst others from Gerhardt of Bonn. The arrangement acts both as a sucking and blowing apparatus, and is furnished with two manometers and proper taps, etc. As I have reason to know that arrangements of this kind work very ill unless really well made, I venture to add that the Gerhardt arrangement to which I refer is No. 239 in his catalogue, and costs about three pounds. It hardly gives enough air, however, to work four blow-pipes, and the blast requires to be steadied by passing the air through a vessel covered with a rubber sheet.

In default of any of these means being available, one of Fletcher's foot-blowers may be employed, but it must be worked very regularly. A table mounted with one blow-pipe made on this plan, and worked by a double-acting bellows, is recommended for students' use. For working flint glass, the air jet may be one-eighth of an inch in diameter and the pressure higher — this will give a brush flame. See Fig. 25.

It will be seen, on looking at the sketch of the blowpipe system, that the pair of blow-pipes farther from the observer can be caused to approach or recede at will by means of a handle working a block on a slide. It often happens that after using all four blow-pipes at once it is necessary to have recourse to one blow-pipe only, and to do this conveniently and quickly is rather an object. Now, in my arrangement I have to turn off both the gas and air from the farther system, and then put in a bit of asbestos board to prevent the nozzles being damaged by the flame or flames kept alight. As I said before, when some experience is gained, glassblowing, becomes a very simple art, and work can be done under circumstances so disadvantageous that they would entirely frustrate the efforts of a beginner. This is not any excuse, however, for recommending inferior arrangements.

Consequently, I say that the pipes leading in gas and air should be all branches of one gas and one air pipe, in so far as the two remote and one proximate blow-pipe are concerned, and these pipes should come up to the table to the right hand of the operator, and should have main taps at that point, each with a handle at least 2 inches long. By this arrangement the operator can instantly turn down all the blow-pipes but one, while, if the inverse operation is required, all the three pipes can be started at once. *[Footnote:* I find, since writing the above, that I have been anticipated in this recommendation by Mr. G. S. Ram, The Incandescent Lamp and its Manufacture, p. 114.*]*

The separate air and gas taps must be left for permanent regulation, and must not be used to turn the supply on or cut it off. In some respects this blow-pipe will be found more easy to manage than an oxygas blow-pipe, for the glass is not so readily brought to the very fluid state, and this will often enable a beginner who proceeds cautiously to do more than he could with the more powerful instrument.

Though I have mentioned glass nozzles for the air supply, there is no difficulty in making nozzles of brass. For this purpose let the end of a brass tube of about one-eighth of an inch diameter be closed by a bit of brass wire previously turned to a section as shown (Fig. 6), and then bored by a drill of the required diameter, say .035 inch. It is most convenient to use too small a drill, and to gradually open the hole by means of that beautiful tool, the watchmaker's "broach." The edges of the jet should be freed from burr by means of a watchmaker's chamfering tool (see Saunier's Watchmaker's Handbook, Tripplin, 1882, p. 232, 342), or by the alternate use of a slip of Kansas stone and the broach.

Fig. 6

The construction of this blow-pipe is so simple, that in case any one wishes to use a brush flame, he can easily produce one simply by changing his air jets to bits of the same size (say one-eighth to one-sixteenth of an inch) tubing, cut off clean. To insure success, the ends of the tubes must be absolutely plane and regular; the slightest inequality makes all the difference in the action of the instrument. If a jet is found to be defective, cut it down a little and try again; a clean-cut end is better than one which has been ground flat on a stone. The end of a tube may, however, be turned in a manner hereafter to be described so as to make an efficient jet. Several trials by cutting will probably have to be made before success is attained. For this kind of jet the air-pressure must be greatly increased, and a large Fletcher's foot-blower or, better still, a small double-action bellows worked with vigour will be found very suitable. A fitting for this auxiliary blow-pipe is shown in Fig. 5 at B.

Professor Roentgen's discovery has recently made it necessary to give more particular attention to the working of soft soda glass, and I have been obliged to supplement the arrangements described by a table especially intended for work with glass of this character. The arrangement has proved so convenient for general work that I give the following particulars. The table measures 5 feet long, 2 feet 11 inches wide, and is 2 feet 9 inches high.

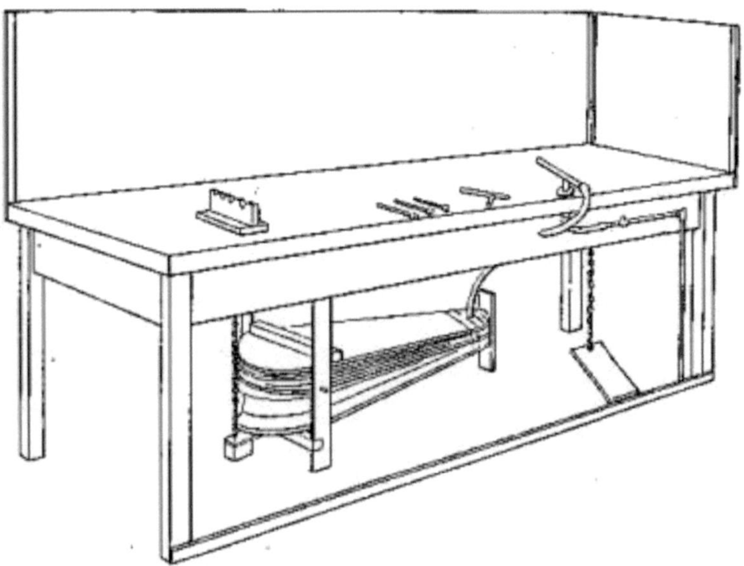

Fig. 7.

It is provided with a single gas socket, into which either a large or small gas tube may be screwed. The larger tube is 5.5 inches long and 0.75 of an inch in diameter. The smaller tube is the same length, and half an inch in diameter. The axis of the larger tube is 3.5 inches above the table at the point of support, and is inclined to the hori-zontal at an angle of 12. The axis of the smaller tube is 2.5 inches above the surface of the table, and is inclined to the horizontal at the same angle as the larger one.

The air jets are simply pieces of glass tube held in position by corks. The gas supply is regulated by a well-bored tap. The air supply is regulated by treading the bellows — no tap is requisite. The bellows employed are ordinary smiths' bellows, measuring 22 inches long by 13 inches wide in the widest part. They are weighted by lead weights, weighing 26 lbs. The treadle is connected to the bellows by a small steel chain, for the length requires to be invariable. As the treadle only acts in forcing air from the lower into the upper chamber of the bellows, a weight of 13 lbs. is hung on to the lower cover, so as to open the bellows automatically.

The air jets which have hitherto been found convenient are:

for the small gas tube

1. a tube 0.12 inch diameter drawn down to a jet of 0.032 inch diameter for small work;
2. plain tubes not drawn down of 0.14 inch, 0.127 inch, and –0.245 inch diameter, and for the large gas tube, plain tubes up to 0.3 inch in diameter.

The table is placed in such a position that the operator sits with his back to a window and has the black calico screen in front of him and to his right. The object of the screen is to protect the workman against draughts. The table is purposely left unscreened to the left of the workman, so that long tubes may be treated.

17. Other appliances which will be required for GLASS-BLOWING are of the simplest character.

(1) Small corks for closing the ends of tubes.

(2) Soft wax — a mixture of bees' wax and resin softened by linseed oil to the proper consistency, easily found by trial, also used for temporarily closing tubes.

(3) A bottle of vaseline for lubricating.

(4) An old biscuit tin filled with asbestos in shreds, and an asbestos towel or cloth for annealing glass after removal from the flame.

As asbestos absorbs moisture, which would defeat its use as an annealing material, it must be dried if necessary.

(5) *A Glass-Cutter's Knife.* — This is best made out of a fine three-cornered file, with the file teeth almost ground out, but not quite; it should be about 2 inches long. After the surface has been ground several times, it may be necessary to reharden the steel. This is best done by heating to a full red and quenching in mercury. The grindstone employed for sharpening the knife should be "quick," so as to leave a rough edge. I have tried many so-called glass knives "made in Germany," but, with one exception, they were nothing like so good as a small French or Sheffield file. In this matter I have the support of Mr. Shenstone's experience.

(6) A wire nail, about 2 inches long, mounted very accurately in a thin cylindrical wooden handle about 5 inches long by one-quarter of an inch diameter, or, better still, a bit of pinion wire 6 inches long, of which 1.5 inches are turned down as far as the cylindrical core., An old dentists' chisel or filling tool is also a very good form of instrument.

(7) A bit of charcoal about 3.5 inches long and 2 wide, and of any thickness, will be found very useful in helping to heat a very large tube. The charcoal block is provided with a stout wire handle, bent in such a manner that the block can be held close above a large glass tube on which the flames impinge. In some cases it is conveniently held by a clip stand. By the use of such a slab of charcoal the temperature obtainable over a large surface can be considerably increased.

I have seen a wine-glass (Venetian sherry-glass) worked on a table with four blow-pipes, such as is here described, with the help of a block of hard wood held over the heated glass, and helping the attainment of a high temperature by its own combustion.

(8) Several retort stands with screw clips.

(9) Some blocks of wood about 5" X 2" X 2" with V-shaped notches cut in from the top.

(10) A strong pair of pliers.

(11) An apparatus for cleaning and drying the breath, when blowing directly by the mouth is not allowable. The apparatus consists of a solid and heavy block of wood supporting a calcium-chloride tube permanently connected with a tube of phosphorus pentoxide divided into compartments by plugs of glass wool. Care should be taken to arrange these tubes so as to occupy the smallest space, and to have the stand particularly stable. The exit tube from the phosphorus pentoxide should be drawn down to form a nozzle, from, say, half an inch to one-eighth of an inch in diameter, so as to easily fit almost any bit of rubber tube. The entry to the calcium chloride should be permanently fitted to about a yard of fine soft rubber tubing, as light as possible. The ends of this tube should terminate in a glass mouthpiece, which should not be too delicate.

As an additional precaution against dust, I sometimes add a tube containing a long plug of glass wool, between the phosphorus pentoxide and the delivery tube, and also a tube containing stick potash on the entry side of the calcium chloride tube, but it may safely be left to individual judgment to determine when these additions require to be made. In practice I always keep the affair set up with these additions. The communication between all the parts should be perfectly free, and the tubes should be nearly filled with reagents, so as to avoid having a large volume of air to compress before a pressure can be got up.

The arrangement will be clear by a reference to Fig. 8, which illustrates the apparatus in use for joining two long tubes. I have tried blowing-bags, etc., but, on the whole, prefer the above arrangement, for, after a time, the skill one acquires in regulating the pressure by blowing by the mouth and lips is such an advantage that it is not to be lightly foregone.

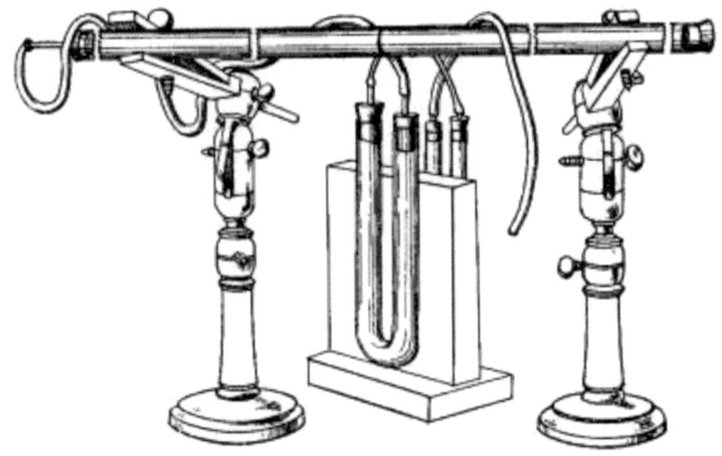

Fig. 8.

18. *The Table.* —

The system of four blow-pipes is, of course, a fixture. In this case the table may be about a yard square, and may be covered with asbestos mill-board neatly laid down, but this is not essential. The table should have a rim running round it about a quarter of an inch high. The tools should be laid to the right of the worker, and for this purpose the blow-pipes are conveniently fixed rather to the left of the centre of the table, but not so far as to make the leg of the table come so close to the operator as to make him uncomfortable, for a cheerful and contented spirit ought to be part of the glass-worker's outfit.

The most convenient height for a blow-pipe table — with the blow-pipes about 2 inches above the table top — is 3 feet 2 inches. Nothing is so convenient to sit upon as a rough music-stool with a good range of adjustment. The advantage of an adjustable seat lies in the fact that for some operations one wants to be well over the work, while in others the advantage of resting the arms against the table is more important.

19. Special Operations. —

The preliminary to most operations before the blow-pipe, is to draw down a tube and pull it out to a fine point. This is also the operation on which a beginner should exercise himself in the first instance. I will suppose that it is desired to draw out a tube about one-quarter of an inch in diameter, with the object of closing it, either permanently or temporarily, and leaving a handle for future operations in the shape of the point, thin enough to cool quickly and so not delay further work.

For this simple operation most of the glass-blower's skill is required. The tube must be grasped between the first finger and thumb of both hands, and held so that the part to be operated on lies evenly between the two hands. The distance between the operator's thumbs may conveniently vary from 2.5 to 4 inches. Releasing the grip of the left hand, let the operator assure himself of his ability to easily rotate the tube about its axis — by the right thumb and finger — he will incidentally observe by the "feel" whether the tube is straight or not.

A good deal of progress can be made from this point before the tube is heated at all. The operator can acquire a habit of instinctively rotating the tube by both hands, however the tube itself be moved about in space, or however it be pushed or pulled. The habit of constant and instinctive rotation is literally about one-third of the whole art of glassblowing. It is unlikely, however, that the beginner will discover that he has not got this habit, until a few failures draw his attention to it.

The glass tube being held in position lightly yet firmly, and the operator being sure that he feels comfortable and at his ease, and that the blow-pipe flame (a single flame in this instance) is well under control, the preliminary heating may be commenced. With a tube of the dimensions given this is a very simple affair. Turn the air partly off, or blow gently, to get a partly luminous gas flame; hold the tube about an inch from the end of this flame, and turn it round and round till it commences to soften.

In the case of soda glass it is usual to employ the gas flame only, but I find that it is better in most cases to use the hot air of a gently-blown flame, rather than have the disadvantage of the soot deposited in the alternative operation. When the glass begins to soften, or even before, it may be moved right into the blow-pipe flame, and the latter may be properly urged.

It is not possible to give quite explicit and definite instructions, applicable to every case, as to when the time is ripe for passing the work into the flame, but the following notes will indicate the general rules to be observed: —

(1) A thick tube must be warmed more slowly and raised to a higher temperature than a thin tube.

(2) The same remark applies to a tube of large diameter, as compared with one of small diameter, whatever the thickness.

(3) In the case of very large or thick tubes the hot air is advantageously employed at first, and to complete the preliminary heating, the luminous flame alone may be used. The object of this is to enable the operator to judge, by the presence of soot, its inability to deposit — or its burning off if deposited — of the temperature of the glass, and of the equality of this temperature all over the surface, for a large and thick tube might be heated quite enough to enable it to be safely exposed to the full heat before it is appreciably yielding to the fingers. In general, when the soot burns off freely, or lead glass begins to show the faintest sign of reduction, or soda glass begins to colour the flame, it is more than safe to proceed.

In order to turn on the full flame the operator will form a habit of holding the work in the left hand only, and he will also take care not to let anything his right hand may be doing cause him to stop rotating the tube with his left thumb and finger.

The preliminary adjustment of air or oxygen supply will enable the change to a flame of maximum power to be made very quickly. The tube having been introduced with constant rotation, it will soon soften sufficiently to be worked. The beginner will find it best to decide the convenient degree of softness by trial.

With soda glass it does not much matter how soft the glass becomes, for it remains viscous, but with lead glass the viscosity persists for a longer time and then suddenly gives place to a much greater degree of fluidity. [Footnote: This is only drawn from my

impressions acquired in glass-working. I have never explicitly tested the matter experimentally.]

It is just at this point that a beginner will probably meet with his
first difficulty. As soon as the glass gets soft he will find that he no
longer rotates the glass at the same speed by the right and left hand,
and, moreover, he will probably unconsciously bend the tube, and
even deform it, by pushing or pulling.

The second third of the art of the glass-blower consists in being
able to move both hands about, rotating a tube with each thumb
and finger, and keeping the distance between the hands, and also
the speed of rotation, constant. Nothing but long practice can give
this facility, but it is essential that it be acquired to some extent, or
no progress can be made. Some people acquire a moderate profi-
ciency very quickly, others, of whom the writer is one, only become
reasonably proficient by months, or even years, of practice.

Supposing that the tube is now ready to be drawn down, the operator will remove it from the flame, and will gently pull the ends
apart, interrupting his turning as little as possible. If the tube be
pulled too hard, or if the area heated be too small (about three-
eighths of an inch in length in the case given would be proper), it
will be found that the ends of the two portions of the tube will be
nearly closed at a very sharp angle (nearly a right angle to the
length of the tube), that the ends will be thin, and that a long length
of very fine tube will be produced. To heat a short length of tube
and pull hard and suddenly is the proper way to make a very fine
capillary tube, but, in general, this is what we want to avoid.

If the operation be successfully performed, the drawn-down tube
will have the appearance exhibited, which is suitable either for subsequently closing or handling by means of the drawn-down portion.
The straightness of the point can be obtained by a little practice in
"feeling" the glass when the tube is rotated as it cools just before it
loses its viscous condition.

When the operation is carried out properly the shoulder of the
"draw" should be perfectly symmetrical and of even thickness, and
its axis regarded as that of a cone should lie in the axis of the tube
produced. The operation should be repeated till the student finds
that he can produce this result with certainty, and he should not be

discouraged if this takes several days, or even weeks. Of course, it is probable that within the first hour he will succeed in making a tolerable job, but it is his business to learn never to make anything else.

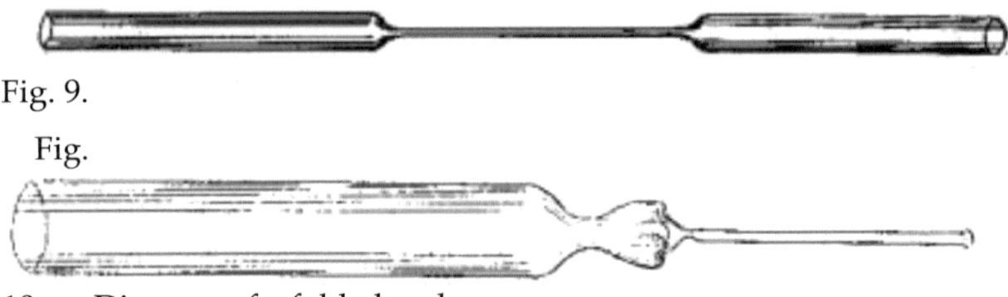

Fig. 9.

Fig.

10. — Diagram of a folded end.

20. Closing and blowing out the End of a Tube. —

When it is desired to close the end of a particular bit of tube, this is easily done by heating the end, and at the same time heating the end of a waste bit of tube or rod; the ends, when placed in contact, stick together, and a point can be drawn down as before. *[Footnote: "Point" is here used in the technical sense, i.e. it is a thin tail of glass produced by drawing down a tube.]* Having got a point, it will be found that the thin glass cools enough to allow of the point being handled after a few moments.

The most convenient way of reducing the point to a suitable length (say 1.5 inch) is to fuse it off in the flame, but this must be done neatly; if a tail is left it may cause inconvenience by catching, or even piercing the finger and breaking off. The blow-pipe flame being turned down to a suitable size, and the shoulder of the "draw" having been kept warm meanwhile, let the tip of the flame impinge on a point where the diameter is about half that of the undrawn tube, and let the temperature be very high (Fig. 11). The tube is to be inclined to the flame so that the latter strikes the shoulder normally, or nearly so. Then, according to circumstances, little or much of the glass can be removed at will by drawing off the tail (Fig. 12), till, finally, a small drop of melted glass only, adheres to the end of the now closed tube (Fig. 13).

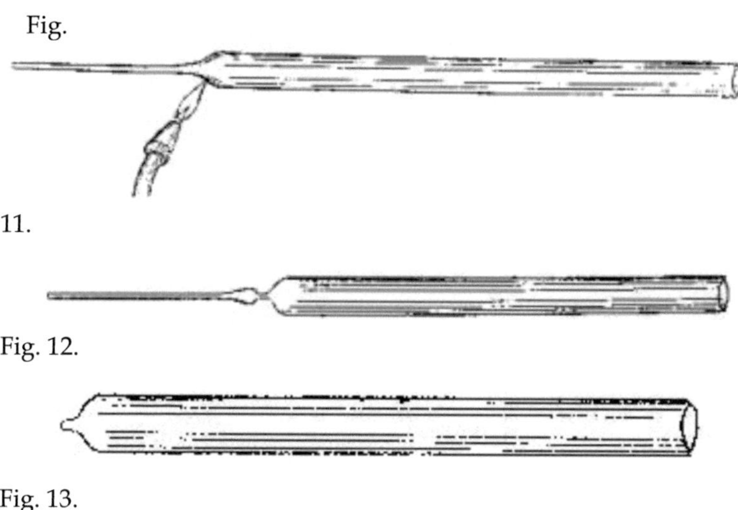

When this is satisfactorily accomplished, heat the extreme end of the tube most carefully and equally, holding it in such a position that the glass will tend to flow from the bead back on to the tube, i.e. hold the closed end up to the flame, the tube being, say, at 45 degrees to the horizontal. Then when the temperature is such as to indicate complete softness lift the tube to the mouth, still holding the tube pointing with its closed end a little above the horizontal, and blow gently. A beginner almost always blows too hard.

What is wanted, of course, is a continued pressure, to give the viscous glass time to yield gradually, if it is uniform; or else intermittent puffs to enable the thinner parts, if there are any, to cool more, and hence become more resisting than the thicker ones. In any case a little practice will enable the operator to blow out a round and even end — neither thicker nor thinner than the rest of the tube.

21. To make a Weld. —

To begin with, try on two bits of glass of the same size, i.e. cut a seven-inch length of glass in half by scratching it with the knife, and pulling the ends apart with a slight inclination away from the scratch. In other words, combine a small bending moment with a considerable tensional stress. It is important to learn to do this properly. If the proportions are not well observed, the tube will break with difficulty, and the section will not be perpendicular to the main length. If the knife is in good order it will make a fine deep scratch — the feel of the glass under the knife will enable the operator to decide when the scratch is made. The operation of cutting large tubes will be treated further on. The two halves of the tube being held one in each hand, and one tube closed at one end, the extremities to be united will be warmed, and then put in the flame as before.

Fig.

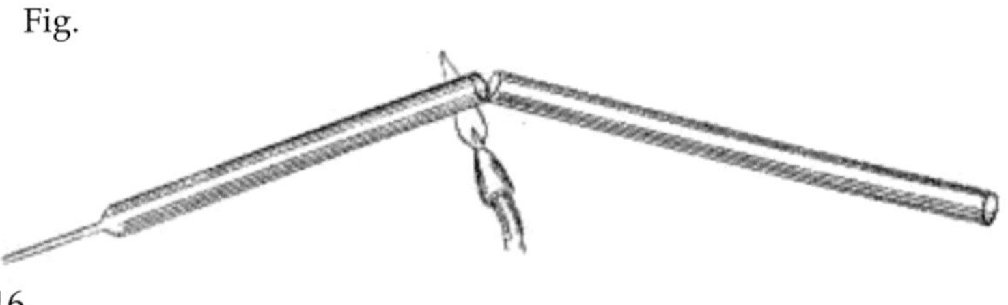

16.

There are many ways of proceeding - perhaps the easiest is as follows. As soon as the glass shows signs of melting at the ends — and care should be taken that much more is not heated — take both bits out of the flame. Stop rotating for a moment, and resting the arms carefully on the edge of the table, raise the tubes above the flame and bring the ends swiftly and accurately together. This is a case of "sudden death no second attempt at making the ends meet can be allowed; if the tubes join in any other than a perfectly exact manner a kink more or less objectionable will result. In practice the operator will learn to bring the ends together, commencing at one point; i.e. the axes of the tubes will be inclined at first, so as to cause adherence at one spot only. If this is not quite "fair," then less damage is done in moving one tube slightly up or down to get the contact exact. The tubes will then be closed upon one another as if they were hinged at the joint. This must be done lightly, yet sufficiently, to ensure that the glass is actually in contact all round.

Having gone so far, replace the tubes — now one — in the flame, and carefully rotating the glass, raise the temperature higher than in the operation just described, in fact the higher the temperature, short of burning the glass, the better. Take the tube out of the flame and blow into the open end, turning constantly as before. One puff is enough. Then turn and pull the glass apart till it is of the same diameter and thickness throughout, and feel that it is straight as before.

Though it is in general of high importance that the joint should be well heated, the beginner will probably find that he "ties up" his glass as soon as it gets really soft.

If his object is to make one joint — at any cost — then let him be careful to use two bits of exactly the same kind of glass, and only get the temperature up to the viscous stage. If the joint be then pulled out till it is comparatively thin, it will probably stand (if of soda glass); certainly, if of lead glass, though in this case it may not be sound. In any case the joint should be annealed in the asbestos box if practicable, otherwise (unless between narrow tubes) with the asbestos rag. Care must be taken that the asbestos is dry.

22. To weld two Tubes of different Sizes. —

To do this, the diameter of the larger tube must be reduced to that of the smaller. The general procedure described in drawing down must be followed, with the following modification. In general, a greater length of the tube must be heated, and it must be made hotter. The tube is to be gradually drawn in the flame with constant turning till the proper diameter and thickness of glass are attained.

Fig.

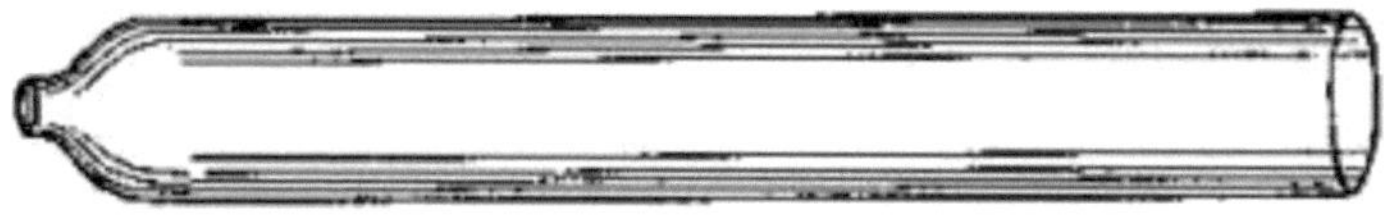

16.

For this operation time must be allowed if the operator's hands are steady enough to permit of it; the shoulder should form partly by the glass sinking in and partly by the process of drawing the hot glass out. A shoulder properly prepared is shown in the sketch. Beginners generally make the neck too thin on large tubes, and too thick on smaller ones. There ought to be no great difference in thickness of glass between the neck on the larger tube, and the smaller tube. The diameters should be as nearly as possible alike.

Having drawn down the larger tube to a neck, take it out of the flame, and as it cools pull and turn till the neck is of the right thickness and is perfectly straight, i.e. make the final adjustment outside the flame, and to that end have the neck rather too thick (as to glass) before it is taken out. It is not necessary to wait till the neck gets cold before the end can be cut off. Make a scratch as before — this will probably slightly damage the temper of the file knife, but that must be put up with. Hold the tube against the edge of the table, so that the scratch is just above the level of the rim, and strike the upper part a smart blow with the handle of the glass knife rather in the direction of its length. *[Footnote:* A bit of hoop iron nailed against the side of the table is a very convenient arrangement, and it need not project appreciably above the general level of the rim.*]*

Of course this applies to a tube where economy has been exercised and the end is short. If the tail is long enough to form a handle, the tube may be pulled apart as before. As a rule a temporary joint between a tube and a rod is not strong enough to enable the shoulder to be broken at the scratch by mere pulling. The ends to be welded must be broken off very clean and true. Subsequent operations are to be carried out as already described.

23. The above operations will be easily performed on tubes up to half an inch in diameter, if they are not too long. It is the length of tube, and consequent difficulty in giving identity of motion with the two hands, which make the jointing of long tubes difficult. There are also difficulties if the tubes are very thin, have a very fine bore or a very large diameter.

All these difficulties merely amuse a good glass-blower, but to an experimenter who wants to get on to other things before sufficient skill is acquired (in the movement of the hands and arms) the following method is recommended. First, use flint glass. Then, assuming that any drawing down has to be done, do it as well as possible, for on this the success of the method to be described especially depends. Be sure that the tubes to be welded are cut off clean and are as nearly as may be of the same size at the point of junction.

To fix the description, suppose it is desired to join two tubes (see Fig. 8), each about one inch in diameter and a yard long. Get four clip stands and place them on a level table. Be sure that the stands are firm and have not warped so as to rock. In each pair of clips place a tube, so that the two tubes are at the same height from the table, and, in fact, exactly abut, with axes in the same straight line. Close one tube by a cork and then fix the blowing apparatus as shown to the other.

In such an operation as this the drying apparatus may be dispensed with, and a rubber tube simply connected to one end of the system and brought to the mouth. Take the oxygen blow-pipe and turn the nozzle till the flame issues towards you, and see that the flame is in order. Then turn down the oxygen till it only suffices to clear the smoky flame, and commence to heat the proposed joint by a current of hot air, moving the flame round the joint. Finally, bring to bear the most powerful flame you can get out of the blow-pipe, and carry it round the joint so quickly that you have the latter all hot at once. Put down the blow-pipe, and, using both hands, press the tubes together (which wooden clips will readily allow), and after seeing that the glass has touched everywhere, pull the tubes a trifle apart. Apply the blow-pipe again, passing lightly over the thin parts, if any, and heating thicker ones; having the end of the rubber tube in his mouth, the operator will be able to blow out thick places. When all is hot, blow out slightly, and having taken the flame away, pull the tubes a little apart, and see that they are straight.

Throw an asbestos rag over the joint, loosen one pair of the clamps slightly, and leave the joint to anneal. It is important that the least possible amount of glass should be heated, hence the necessity of having the ends well prepared, and it is also important that the work should be done quickly; otherwise glass will flow from the upper side downwards and no strong joint will be obtained.

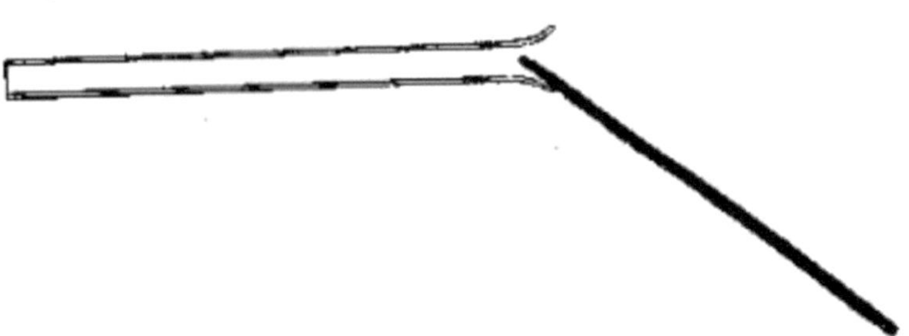

17. — Tube being opened at one end.

24. To weld Tubes of very small Bore.

—

If the bore is not so small as to prevent the entrance of the point of the iron nail, get the ends of the tubes hot, and open the bore by inserting the end of the nail previously smeared over with a trace of vaseline. Work the nail round by holding the handle between the thumb and first finger of the right hand, the tube being similarly placed in the left. The tube and nail should be inclined as shown in the sketch.

Never try to force the operation; the nail soon cools the glass, so that only a very short time is available after each heat; during this the tube should be rotated against the nail rather than the nail against the tube. Be careful not to heat a greater length of tube than is necessary, or the nail will, by its component of pressure along the tube, cause the latter to "jump up" or thicken and bulge. Both ends being prepared, and if possible, kept hot, the weld may be made as before, and the heating continued till the glass falls in to about its previous thickness, leaving a bore only slightly greater than before.

It is in operations such as this that the asbestos box will be found of great use. As soon as one end of the weld is ready cool it in the flame till soot deposits, and then plunge it into the asbestos. This will cause it to cool very slowly, and renders it less likely to crack when again brought into the flame. Turned-out ends, if the glass is at all thick, are very liable to crack off on reheating, so that they must be reintroduced (into the flame) with especial care. This liability to breakage is reduced, but not eliminated, by the asbestos annealing.

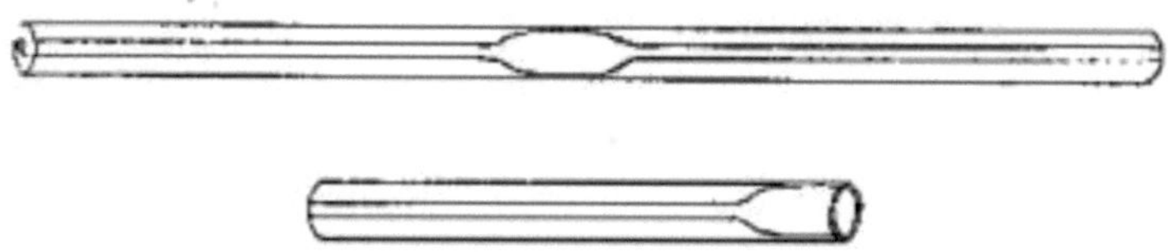

Figs. 18 and 19.

25. When the bore is very fine, it is best to seal off the tubes, and blow an incipient bulb near one end of each tube. These bulbs may be cooled in asbestos, and cut across when cold by means of a scratch touched at one end (Figs. 18 and 19) by a fine point of highly incandescent glass. For details of this method see p. 46, Fig. 21. Time is occasionally saved by blowing off the ends of the bulbs. The details of this process will be described when the operation of making thistle-headed tubes is dealt with.

26. When the tubes are both of large diameter, long, and very thin (cylinder tubes), a considerable amount of difficulty will be experienced. On the whole, it is best to heat each end separately till the glass thickens a little, anneal in the flame and in asbestos, and then proceed as in 22. If the ends are not quite true, it will be found that quite a thickness of glass may be "jumped" together at one side of the tubes, while the edges are still apart at the other. When this looks likely to happen, incline the tubes as if the joint were a hinge, and bend back quickly; do not simply continue to push the tubes together in a straight line, or an unmanageable lump of glass will be formed on one side.

If in spite of these precautions such a lump does form, proceed as follows. Take a rod of glass, at least one-eighth of an inch thick, and warm it in the flame at one end. Heat the imperfect joint till it softens all round, and then bring the flame right up to the thick part, and heat that as rapidly and locally as possible. The oxygas flame does this magnificently. Press the heated end of the glass rod against the thick part, and pull off as much of the lump as it is desired to remove, afterwards blowing the dint out by a judicious puff. Finish off as before.

27. Occasionally, when it is seen that in order to produce a joint closed all round, one side of the tube would be too much thickened, it is better to patch the open side. For this purpose take a glass rod about one-sixteenth inch in diameter, and turn the flame to give its greatest effect, still keeping rather an excess of air or oxygen. See that the side of the joint already made is kept fairly hot — it need not be soft; interrupt any other work often enough to ensure this. Then, directing the flame chiefly on the thin rod, begin to melt and pull the glass over the edges of the gap. When the gap is closed get the lump very hot, so that all the glass is well melted together, and then, if necessary, pull the excess of glass off, as before described.

It must be remembered that this and the method of the previous section are emergency methods, and never give such nice joints as a manipulation which avoids them, i.e. when the ends of the tubes are perfectly straight and true to begin with. Also note that, as the tubes cannot be kept in rotation while being patched, it is as well to work at as low a temperature as possible, consistently with the other conditions, or the glass will tend to run down and form a drop, leaving a correspondingly thin place behind.

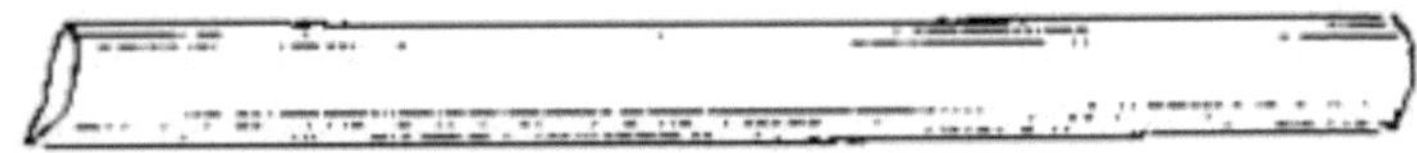

Fig. 20.

28. A very common fault in cutting a tube of about an inch in diameter is to leave it with a projecting point, as shown. This can be slowly chipped off by the pliers, using the jaws to crush and grind away the edge of the projection; it is fatal to attempt to break off large pieces of glass all at once.

29. It will be convenient here to mention some methods of cutting large tubes. With tubes up to an inch and a half in diameter, and even over this — provided that the glass is not very thick — we may proceed as follows: Make a good scratch about half an inch long, and pretty deep, i.e. pass the knife backwards and forwards two or three times. Press a point of melted glass exactly on one end of the scratch; the glass point even when pressed out of shape should not be as large as a button one-twelfth of an inch in diameter. If this fails at first, repeat the operation two or three times.

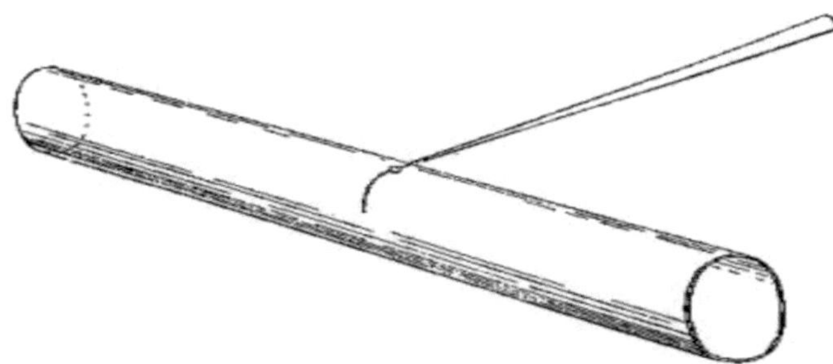

Fig. 21.

If a crack does not form, touch the hot place with the cold end of the nail. If no success is obtained, try the other end of the scratch. If failure still pursues the operator, let him make another cut on the opposite side of the tube and try again. In general, the tube will yield the first or second time the hot drop of glass is applied. Never apply the drop at the centre of the scratch, or a ragged crack, which may run in any direction, will result. Very often, with a large tube, the crack formed by a successful operation will only extend a short distance. In this case it is desirable to entice the crack round the tube, and not trust to its running straight when the tube is pulled apart.

On the whole, the best method in this case is to employ a flame pencil, which should be kept ready for use. This merely consists of a bit of glass tube of about the same dimensions as an ordinary lead pencil, drawn down to a very fine jet at one end. The jet must not be very long or thin, or the glass will soon fuse up. A few trials will enable the operator to get the proper proportions, which are such that the tube has the general appearance of a pencil normally sharpened (say with a cone of 60'). This tube is best made of hard glass. Connect it to a gas supply by light flexible tubing, and turn down the gas till the flame from the end of the jet is not more than one-tenth of an inch long. Then apply the jet, beginning from the end of the crack, and gradually draw it (the crack) round the tube. The operation will be assisted if a rubber ring is slipped on the tube to begin with, so that the eye has some guide as to whether the flame is being drawn round properly or not. The ring must, of course, be far enough away to escape the effect of the flame. The crack will be found to follow the flame in the most docile manner, unless the tube is thick or badly annealed. Some operators recommend a pencil of glowing charcoal, but the flame is undoubtedly better.

30. To cut very thick Tubes. -

A large number of methods have been proposed, and nearly everybody has his favourite. The following has always succeeded with me. First mark on the tube, by means of a little dead black spirit paint, exactly where the cut is to be. Then sharpen the glass knife and scratch a quite deep cut all round: there is no difficulty in making the cut one-twentieth of an inch deep. It will be proper to lubricate the knife with kerosene after the first mark is made. *[Footnote: The edge of the knife may be advantageously saved by using an old file moistened with kerosene for the purpose. I find kerosene is not worse, but, if anything, better than the solution of camphor in turpentine recommended by Mr. Shenstone.]*

If the glass is about one-eighth of an inch thick, the scratch maybe conveniently about one-twentieth of an inch deep, but if the glass is anything like one-quarter of an inch thick, the scratch must be much deeper, in fact, the glass may be half cut through. To make a very deep scratch, a wheel armed with diamond dust, which will be described later on, may be used. However, it is not essential to use a diamond wheel, though it saves time.

When the cut is made to a sufficient depth proceed thus: Obtain two strips of bibulous paper or bits of tape and twist them round the tube on each side of the scratch, allowing not more than one-eighth of an inch between them. Then add a few drops of water to each, till it is thoroughly soaked, but not allowing water to run away. Dry out the scratch by a shred of blotting paper.

Turn down the oxygas flame to the smallest dimensions, and then boldly apply it with its hottest part playing right into the nick and at a single point. Probably in about two seconds, or less, the tube will break. If it does not, rotate the tube, but still so that the flame plays in the nick. After making the tube very hot all round — if it has not broken — apply the flame again steadily at one point for a few seconds and then apply a bit of cold iron. If the tube does not break at once during these processes, let it cool, and cut the groove deeper;

then try again. *[Footnote: This method is continually being reinvented and published in the various journals. It is of unknown antiquity.]*

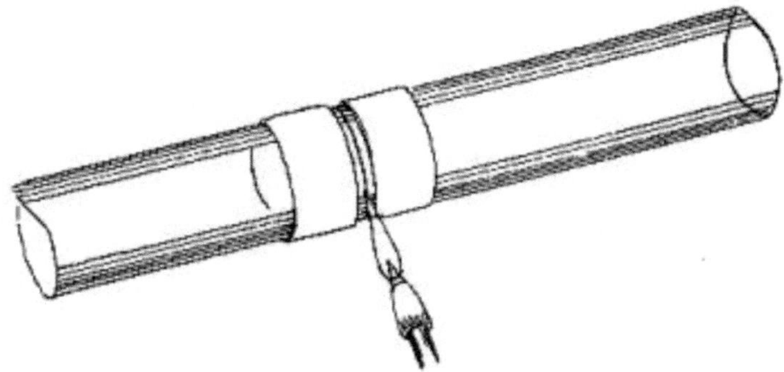

Fig. 22.

If the tube breaks after great heating and long efforts, it will probably leave incipient cracks running away from the break, or may even break irregularly. A good break is nearly always one that was easily made. If a number of rings have to be cut, or a number of cuts made on glass tubes of about the same size, it will be found economical in the end to mount a glazier's diamond for the purpose. A simple but suitable apparatus is figured (Fig. 23).

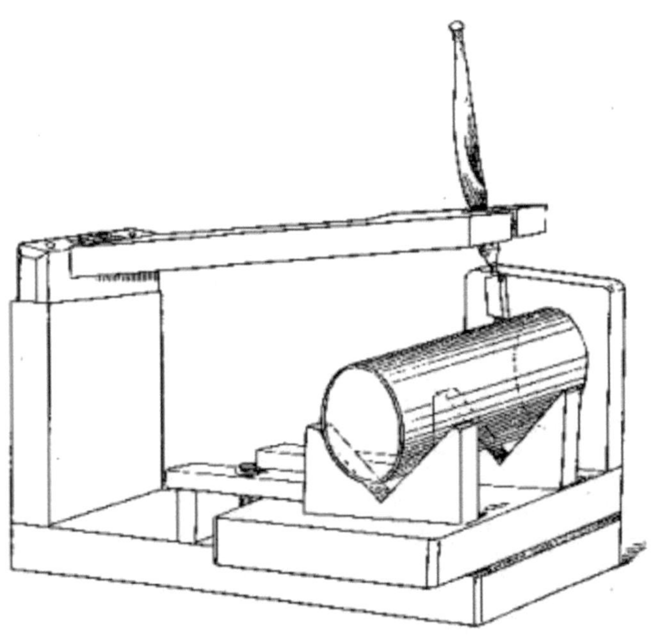

Fig. 23.

The only difficulty is to regulate the position of the diamond so that it cuts. In order to do this, carefully note its cutting angle by preliminary trials on sheet glass, and then adjust the diamond by clamps, or by wriggling it in a fork, as shown. Weight the board very slightly, so as to give the small necessary pressure, and produce the cut by rotating the tube by hand. When a cut is nearly completed take great care that the two ends join, or irregularity will result. This is not always easy to do unless the tube happens to be straight. Having got a cut, start a crack by means of a fine light watchmaker's hammer, or even a bit of fused glass, and entice the crack round the cut by tapping with the hammer or by means of the flame pencil.

If the cut is a true "cut" the tube will break at once. As a supply of electrical current for lighting will, in the near future, be as much a matter of course for laboratory purposes as a gas supply, I add the following note. To heat a tube round a scratch, nothing — not even the oxygas blow-pipe — is so good as a bit of platinum or iron wire electrically heated. If the crack does not start by considerable heating of the glass, stop the current, unwind the wire, and touch the glass on the crack either with a bit of cold copper wire or a wet match stem. I prefer the copper wire, for in my experience the water will occasionally produce an explosion of cracks. On the other hand, the cold wire frequently fails to start a crack.

Judging from the appearance of thick tubes as supplied by the dealers, the factory method of cutting off appears to be to grind a nick almost through the tube, and right round; and for really thick glass this is the safest but slowest way; a thin emery wheel kept wet will do this perfectly. Suitable wheels may be purchased from the "Norton" Emery Wheel Co. of Bedford, Mass., U.S.A. — in England through Messrs. Churchill and Co. of London, importers.

31. To blow a Bulb at the End of a Tube. —

I must admit at once that this is a difficult operation — at all events, if a large bulb is required. However, all there is to be said can be said in few words. In general, when a bulb is required at the end of a tube it will be necessary to thicken up the glass. A professional glass-worker will generally accomplish this by "jumping up" the tube, i.e. by heating it where the bulb is required, and compressing it little by little until a sufficient amount of glass is collected. The amateur will probably find that he gets a very irregular mass in this way, and will be tempted to begin by welding on a short bit of wide and thick tubing preparatory to blowing out the bulb.

However, supposing that enough glass is assembled by either of these methods, and that it is quite uniform in thickness, let the thickened part be heated along a circle till it becomes moderately soft, and let it then be expanded about one-fifth, say by gently blowing. It is perhaps more important to keep turning the glass during bulb-blowing than in any other operation, and this both when the glass is in the flame and while the bulb is being blown. It is also very important to avoid draughts. In general, a bulb is best blown with the tube in a nearly horizontal position, but sloping slightly upwards from the mouth. If it be noticed that a bulb tends to blow out more at one side than another, let the side of greatest protuberance be turned down, so that it is at the lowest point, reduce the pressure for an instant, and then blow again. It will be observed that the bulb will now expand at the top.

The reason of this is chiefly that the under side cools most rapidly (according to Faraday, *Chemical Manipulation*, 1194), and consequently can expand no further; but also it is not unlikely that the glass tends to flow somewhat from the upper side, which remains hot, and consequently the bulb, when the next puff reaches it, will tend to yield at this point. By heating several zones the tube will become gradually expanded.

Fig.

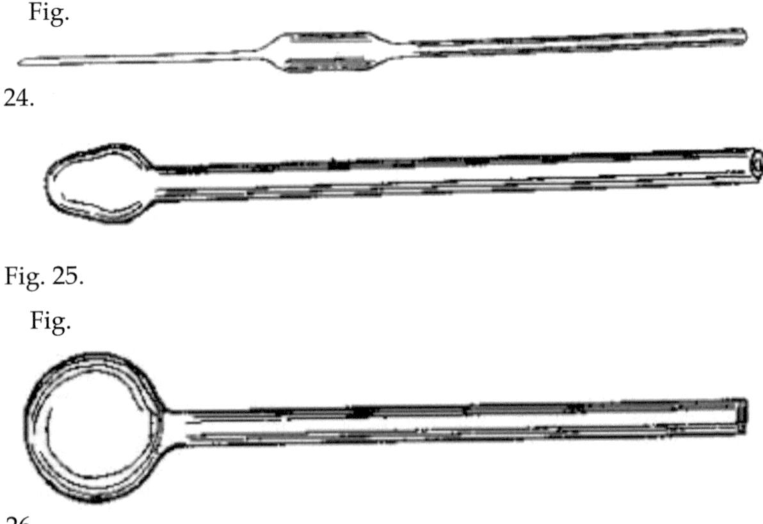

24.

Fig. 25.

Fig.

26.

When the length of the thickened part of the tube only slightly exceeds its diameter (Fig. 25), let the whole be brought to a temperature which, with flint glass, should be just short of that of perfect fluidity; and then, holding the tube horizontally and constantly turning it, let the bulb be blown out to its full size, noting the appearances and correcting too great protuberance on any side by the means above mentioned. If the bulb appears pear-shaped turn the tube so that the melted mass is directed upwards; if the bulb have the contrary fault, correct in the corresponding manner.

The bulb when finished may be lightly tapped on the table, when, if there is any weak place owing to inequality of thickness, the bulb will break, and the operation may be started afresh. "A good bulb is round, set truly on the tube, and free from lumps of thick glass or places of excessive thinness." When the amateur has succeeded in blowing a bulb two inches in diameter on the end of a strong bit of thermometer tube — say for an air thermometer — he may well seek the congratulations of his friends.

In case the bulb is not satisfactory on a first attempt, it may be melted down again, if the following precautions are taken. Directly creases begin to appear in the bulb let it be withdrawn from the flame, and gently blown till the creases come out. By alternate heating and blowing the glass can be got back to its original form, or nearly so, but unless the operator shows great skill and judgment, the probability is that the glass will be uneven. By heating and keeping the thicker parts in the higher position, and blowing a little now and again, the glass may be got even, and a new attempt may be made. It must not be supposed that this process can be carried on indefinitely, for the glass tends to lose its viscous properties after a time, or, at all events, it "perishes" in some way, especially if it has been allowed to get very thin; consequently too frequent attempts on the same glass are unprofitable. Two or three trials are as many as it generally pays to make. As a rule the largest possible flame may be used with advantage in this operation.

32. To blow a bulb in the middle of a tube,

the procedure is much like that already treated, but the manipulation is, if anything, more difficult, for the further end of the tube must be carried and turned as well as the end which is held to the lips.

33. To make a side Weld. —

This is by no means difficult, but is easier with lead glass than with soda glass. The tube to which it is desired to make a side connection having been selected, it is closed at one end by rubber tube stops, or in any other suitable manner. The zone of the proposed connection is noted, and the tube is brought to near softness round that circle (if the tube is made actually soft, inconvenience will arise from the bending, which is sure to occur). Two courses are then open to the operator, one suitable to a thick tube, the other to a tube of moderate thickness.

Taking the former first. Provide a piece of glass rod and warm its end. Direct a small flame against the spot on the thick tube where the proposed joint is to be. When the glass becomes almost incandescent at this spot, put the end of the rod against it and draw out a thread of glass till sufficient "metal" has been removed. Then fuse off the thread close to the tube.

Fig. 27.

The subsequent procedure is the same as for thin tubes. In this case heat the spot by the smallest flame available, and get the spot very hot. Blow it out gently into a bubble, perhaps extending to a height equal to its diameter. Then heat the top of the bubble till it is incandescent and blow violently. This will produce an opening fringed by glass so thin as to exhibit interference colours. Remove the filmy part, and heat the frayed edges till they cohere and form an incipient tube. If the flame has been of a correct size, the tube will now be of the same diameter as the tube to be welded on, and will project perhaps one-sixteenth of an inch from the surface of the main tube (Fig. 28).

Fig. 28.

Fig.

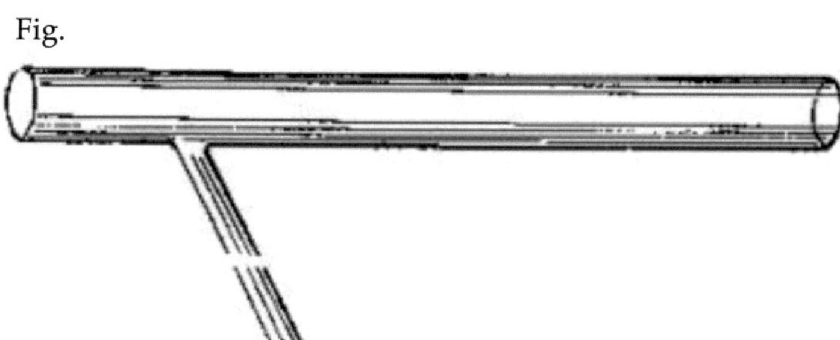

29.

When this stage is reached, again heat the tube all round till it nearly softens, and by means of the other hand heat the end of the other tube which it is proposed to weld. Just before the main tube actually softens, turn it so as to heat the edges of the aperture, and at the same time get the end of the side tube very hot. Take both out of the flame for an instant, and press the parts together, instantly slightly withdrawing the side tube. If the operation is well performed, it will be found that the point of maximum thickness of glass is now clear of the main tube. The joint is then to be heated all round and blown out — a rather awkward operation, and one requiring some practice, but it can be done.

Fig.

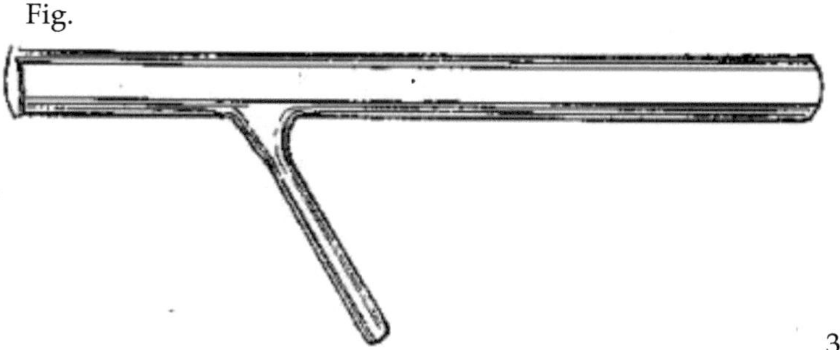

30.

If great strength is wanted, heat the main tube all round the joint bit by bit, and blow each section slightly outwards. If the operator is confident in his skill, he should then heat the whole joint to the softening point, blow it out slightly, and then adjust by pulling and pushing. Cool first in the gas flame, and then plunge the joint into the asbestos and cover it up — or if too large, throw the asbestos cloth round it.

In the case of soda glass this final "general heat" is almost essential, but it is not so with flint glass, and as the general heat is the most difficult part of the job, it will be found easier to use lead glass and omit the general heating. With soda glass a very small irregularity will cause the joint to break when cold, but flint glass is much more long-suffering. It is easy to perform the above operation on small tubes. For large ones it will be found best to employ flint glass and use the clip stands as in the case of direct welds, treated above, but, of course, with suitable modifications. Never let the main tube cool after the hole is made until the work is done.

34. Inserted Joints. —

In many instances the performance of apparatus is much improved by joints of this kind, even when their use is not absolutely essential.

There are two ways in which inserted joints may be made. The first method is the easier, and works well with flint glass; but when one comes to apply it to soda glass there is a danger of the glass becoming too thick near the joint, and this often leads to a cracking of the joint as the glass cools.

Fig. 31.

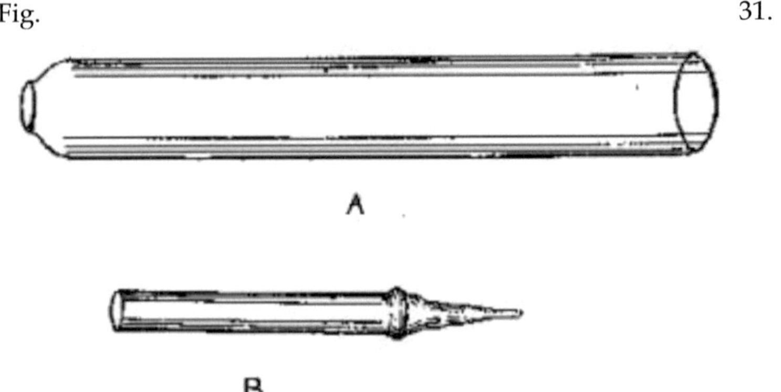

Suppose it is desired to insert the tube B into the tube A (Fig. 31). Begin by reducing the size of the end of tube A till B will just slip in quite easily. With B about one-quarter inch in diameter, a clearance of about one-twentieth of an inch, or less, in all (i.e. one-fortieth of an inch on each side) will be proper.

Heat B by itself at the proposed zone of junction, and blow out a very narrow ring; then compress this slightly so that it forms an almost closed ring of glass. The figure refers to the close of this operation (Fig. 31, B). It does not matter much whether the ring remains a mere flattened bulb, or whether it is a solid ring, but it must be one or the other. Some judgment must be exercised in preparing the ring. In general, the beginner will collect too much glass in the ring, and consequently the joint, when made, will either be thick and liable to crack easily, or it will be blown out into an erratic shape in endeavours to reduce this thickness. Accordingly, the operator will, if necessary, thin the tube B by drawing slightly, if he considers it desirable, before the little enlargement is blown out. In general, two heats must be used for this operation.

Fig.

32.

Get the approximating parts of both A and B up to a temperature just below that at which they will adhere, and having closed the other end of A, place B carefully within it up to the ring, and if it can be arranged, have a mica wad in A, with a central hole through which the end of B can project. This will very much facilitate the operation, especially if B is long, but may be dispensed with by the exercise of care and skill.

The operation is now simple. Fuse the junction and press the tubes lightly together, being careful not to collect more glass than can be helped; finally, blow out the joint and reduce the thickness by mild drawing (Fig. 33). In order to make a really good joint, two points must be particularly attended to — the rim must be thin and its plane perfectly perpendicular to the axis of tube B; the end of tube A must be cut off quite clean and perpendicular to its axis before B is inserted. So important are these conditions — especially the latter that the writer has even occasionally used the grindstone to get the end of A into a proper condition, an admission which will probably earn the contempt of the expert glass-worker.

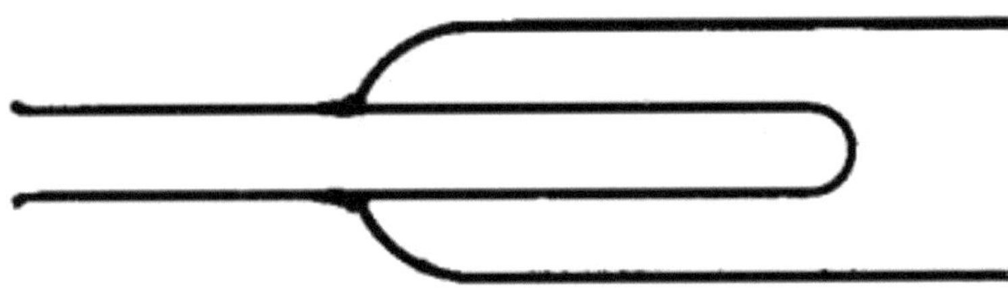

Fig. 33.

Now for the second method, which is often practised in Germany, where soda glass is chiefly used. With this glass the chief point is to get a very even and not too thick ring at the junction, and consequently the extra thickening produced by making a rim on B is rather a drawback. The method consists in cutting off from B the length which it is desired to insert, slipping this into A (which may be an otherwise closed bulb, for instance), and then gradually melting up the open end of A till the piece of B inside will no longer fall out. By holding the joint downwards so that the inserted portion of B rests on the edges of the opening, a joint may be made with the minimum thickening.

The external part of B, previously heated, is then applied, and the joint subjected to a "general" heat and blown out. Very nice joints may be made by this method, and it is perhaps the better one where the external part of B is to be less in diameter than the inserted part. It was in this manner that the writer was taught to make glass velocity pumps, one of which, of a good design, is figured as an example.

In all cases good annealing should follow this operation. If the inserted part of the inner tube (B) is anything like an inch in diameter, and especially if it is of any length, as in some forms of ozone apparatus, or in a large Bunsen's ice calorimeter, the arrangements for supporting the inner part must be very good. A convenient way of proceeding when the inner tube is well supported is to make the mouth of A only very little larger than the diameter of B, so that B will only just slip in. Then the mouth of A and the zone of B may be heated together, and B blown out upon A. This, of course, must be arranged for, if necessary, by temporarily stopping the inner end of B.

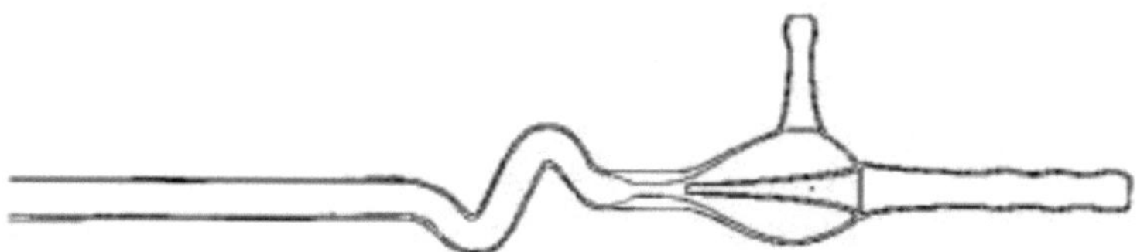

The inner support of B should be removed as soon as practicable after the joint is made, or, at all events, should not be perfectly rigid; a tightly-fitting cork, for instance, is too rigid. The reason is, of course, that in cooling there may be a tendency to set B a little to one side or the other, and if it is not free to take such a set, the joint most probably will give way. Good annealing both with flame and asbestos is a *sine qua non* in all inserted work.

Fig. 34.

35. Bending Tubes. —

I have hitherto said nothing about bending tubes, for to bend a tube of a quarter of an inch in diameter, and of ordinary thickness, is about the first thing one learns in any laboratory, while to bend large tubes nicely is as difficult an operation as the practice of GLASS-BLOWING affords. However, even in bending a narrow tube it is possible to proceed in the wrong way. The wrong way is to heat a short length of the tube and then bend it rapidly, holding the plane of the bend horizontal. The right way, *per contra,* is to use a batswing burner to heat, say, two inches of the tube with constant turning till it is very soft, and then, holding the glass so that the bend will be in a vertical plane passing through one eye (the other being shut), to make the bend rather slowly.

If an exact angle is required, it is as well to have it drawn out on a sheet of asbestos board. In this case bend the glass as described till it is approximately right, and finish by laying it on the asbestos board and bringing it up to the marks. A suitable bit of wood may be substituted for the asbestos on occasion.

N.B. — The laboratory table is not a suitable piece of wood. A right-angled bend is often wanted. In this case the corner of a table will serve as a good guide to the eye, the glass being finished by being held just above it. If great accuracy is wanted, make a wooden template and suspend it by a screw from the side of the table, so that the vertex of the gauge for the interior angle projects downwards, then finish by bending the tube round it. The wood may be about half an inch thick.

If a sharp bend is required, heat the tube in the blow-pipe, and bend it rapidly, blowing out the glass meanwhile. The reason why a long bend should be held in a vertical plane is that the hot part tends to droop out of the plane of the bend if the latter be made in a horizontal position. To bend a tube above half an inch in diameter is a more or less difficult operation, and one which increases in difficulty as the diameter of the tube increases.

A U-tube, for instance, may be made as follows: Use the four blow-pipe arrangement so as to heat a fair length of tube, and get, say, two inches of tube very hot--almost fluid, in fact — by means of the carbon block supported from a stand. Remove the tube rapidly from the flame and draw the hot part out to, say, three inches. Then, holding the tube so as to make the bend in a vertical plane, bend it and blow it out together to its proper size.

This operation seems to present no difficulties to experienced glass-workers, even with tubes of about one inch in diameter, but to the amateur it is very difficult. I always look on a large U-tube with feelings of envy and admiration, which the complex trick work of an elaborate vacuum tube does not excite in the least. It will be noted that this method may, and often does, involve a preliminary thickening of the glass.

With tubes over an inch in diameter I have no idea as to what is the best mode of procedure — whether, for instance, a quantity of sand or gas coke might not be used to stuff out the tube during bending, but in this case there would be the difficulty of removing the fragments, which would be sure to stick to the glass.

Of course, if the bend need not be short, the tube could be softened in a tube furnace and bent in a kind of way. I must admit that with tubes of even less than one inch in diameter I have generally managed best by proceeding little by little. I heat as much of the glass as I can by means of a gigantic blow-pipe, having a nozzle of about an inch in diameter, and driven by a machine-blower.

When I find that, in spite of blowing, the tube begins to collapse, I suspend operations, reheat the tube a little farther on, and so proceed. If by any chance any reader knows a good laboratory method of performing this operation, I hope he will communicate it to me. After all, the difficulty chiefly arises from laboratory heating appliances being as a rule too limited in scope for such work.

The bending of very thin tubes also is a difficulty. I have only succeeded here by making very wide bends, but of course the blowing method is quite applicable to this case, and the effect may be obtained by welding in a rather thicker bit of tube, and drawing and blowing it till it is of the necessary thinness. This is, however, a mere evasion of the difficulty.

36. Spiral Tubes. —

These are easily made where good heating apparatus is available. As, however, one constantly requires to bend tubes of about one-eighth inch in diameter into spirals in order to make spring connections for continuous glass apparatus, I will describe a method by which this is easily done. Provide a bit of iron pipe about an inch and a quarter in outside diameter. Cover this with a thick sheath of asbestos cloth, and sew the edges with iron wire. Hammer the wire down so that a good cylindrical surface is obtained. Make two wooden plugs for the ends of the iron pipe. Bore one to fit a nail, which may be held in a small retort clip, and fasten a stout wire crank handle into the other one. Support the neck of the handle by means of a second clip. In this way we easily get a sort of windlass quite strong enough for our purpose.

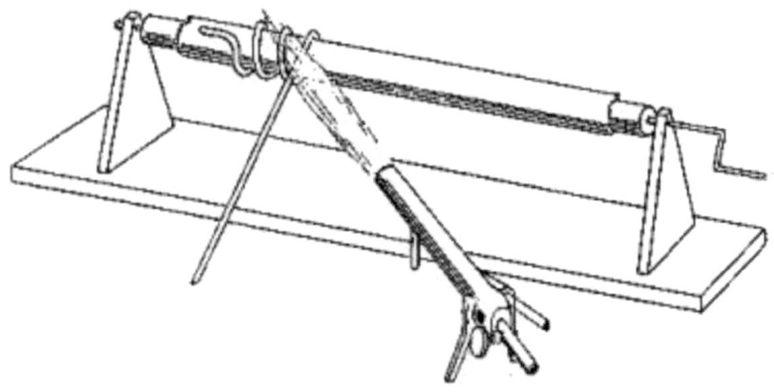

35.

Provide a large blow-pipe, such as the blow-pipe of a Fletcher crucible furnace, Select a length of tubing and clean it. Lash one end to the cylinder by means of a bit of wire, and hold the other end out nearly horizontally. Then start the blow-pipe to play on the tube just where it runs on to the asbestos cylinder, and at first right up to the lashing. Get an attendant to assist in turning the handle of the wind-lass, always keeping his eye on the tube, and never turning so fast as to tilt the tube upwards. By means of the blow-pipe, which may be moved round the tubing, heat the latter continuously as it is drawn through the flame, and lay it on the cylinder in even spirals.

If the tubing is thin, a good deal of care will have to be exercised in order to prevent a collapse. A better arrangement, which, however, I have not yet tried, would, I think, be to replace the blow-pipe by two bats-wing burners, permanently fastened to a stand, one of them playing its flame downwards on to the top of the flame of the other. The angle between the directions of the jets might be, say, 130, or whatever is found convenient. In this way the glass would not be so likely to get overheated in spots, and better work would doubtless result. However, I have made numbers of perfectly satis-factory spirals as described. Three or four turns only make a suffi-ciently springy connection for nearly all purposes.

37. On Auxiliary Operations on Glass:-

Boring Holes through Glass. — This is much more easily done than is generally supposed. The best mode of procedure depends on the circumstances. The following three cases will be considered:-

1. Boring holes up to one-quarter inch diameter through thick glass (say over one-eighth inch), or rather larger holes through thin glass.

2. Boring holes of any size through thick glass.

3. Boring round holes through ordinary window glass.

38. Boring small Holes. —

Take a three-cornered file of appropriate dimensions, and snip the point off by means of a hammer; grind out most of the file marks to get sharp corners. Dip the file in kerosene, and have plenty of kerosene at hand in a small pot. Place the broken end of the file against the glass, and with considerable pressure begin to rotate it (the file) backwards and forwards with the fingers, very much as one would operate a bradawl against a hard piece of wood. The surface of the glass will shortly be ground away, and then the file bradawl will make much quicker progress than might be expected. Two or three minutes should suffice to bore a bit of sheet window-glass.

The following points require attention:

(1) Use any quantity of oil.

(2) After getting through the skin reduce the pressure on the file.

(3) Be sure to turn the file backwards and forwards through a complete revolution at least.

(4) When the hole is nearly through reduce the pressure.

(5) When the hole is through the glass be exceedingly careful not to force the file through too rapidly, otherwise it will simply act as a wedge and cause a complete fracture.

(6) In many cases it is better to harden the file in mercury before commencing operations; both files and glass differ so much in hardness that this point can only be decided by a trial. If it is found

necessary to harden the file, use either a large blow-pipe and a coke or charcoal bed, or else a small forge. A small blowpipe, such as is generally found in laboratories, does more harm than good, either by burning the end of the file or raising it to an insufficient temperature.

(7) To sharpen the file, which is often necessary after passing through the "skin" of the glass, put it in a vice so that the point just protrudes clear of the jaws. Then, using a bit of waste iron as an intermediary anvil or punch, knock off the least bit from the point, so as to expose a fresh natural surface. The same result may be brought about by the use of a pair of pliers.

If several holes have to be bored, it is convenient to mount the file in the lathe and use a bit of flat hard wood to press up the glass by means of the back rest. A drilling machine, if not too heavy, does very well, and has the advantage of allowing the glass to remain horizontal so that plenty of oil can be kept in the hole.

Use a very slow speed in either case — much slower than would be used for drilling wrought iron. It is essential that the lubricant should flow on to the end of the file very freely, either from a pipette or from the regular oil-feed. If a little chipping where the file pierces the back surface is inadmissible, it is better, on the whole, to finish the bore by hand, using a very taper file. It is not necessary to use a special file for the lathe, for a well-handled file can be chucked very conveniently in a three-jaw chuck by means of the handle.

Mr. Shenstone recommends a lubricant composed of camphor dissolved in turpentine for general purposes. With the object of obtaining some decisive information as to the use of this lubricant, and to settle other points, I made the following experiments. Using an old three-cornered French file, I chipped off the point and adjusted the handle carefully. I also ground out the file marks near the point, without hardening the file in mercury. Using kerosene and turpentine and camphor, I began to bore holes in a hard bit of 3/32 inch window glass.

Each hole was bored to about one-eighth inch in diameter in four minutes with either lubricant. After hardening the file in mercury and using kerosene, I also required four minutes per hole. After mounting the file in a lathe which had been speeded to turn up brass rods of about one-half-inch diameter, and therefore ran too fast, I required one and a half minutes per hole, and bored them right through, using kerosene. On the whole, I think kerosene does as well as anything, and for filing is, I think, better than the camphor solution. However, I ought to say that the camphor-turpentine compound has probably a good deal to recommend it, for it has survived from long ago. My assistant tells me he has seen his grandfather use it when filing glass.

I beg to acknowledge my indebtedness to Mr. Pye, of the Cambridge Scientific Instrument Company, for showing me in 1886 (by the courtesy of the Company) the file method of glass-boring; it is also described by Faraday in *Chemical Manipulation*, 1228.

It is not necessary, however, to use a file at all, for the twist drills made by the Morse Drill Company are quite hard enough in their natural state to bore glass. The circumferential speed of the drill should not much exceed 10 feet per minute. In this way the author has bored holes through glass an inch thick without any trouble except that of keeping the lubricant sufficiently supplied. For boring very small holes watchmaker's drills may be used perfectly well, especially those tempered for boring hardened steel. The only difficulty is in obtaining a sufficient supply of the lubricant, and to secure this the drill must be frequently withdrawn.

My reason for describing the file method at such length is to be found in the fact that a Morse drill requires to be sharpened after drilling glass before it can be used in the ordinary way, and this is often a difficulty.

I ought to say that I have never succeeded in boring the *barrel of a glass tap* by either of these methods. [Footnote: I have been lately informed that it is usual to employ a splinter of diamond set in a steel wire holder both for tap boring and for drilling earthenware for riveting. The diamond must, of course, be set so as to give sufficient clearance for the wire holder.

For methods of using and setting diamond tools see 55. It will suffice to say here that a steel wire is softened and filed at one end so as to form a fork; into this the diamond is set by squeezing with pliers. The diamond is arranged so as to present a point in the axis of the wire, and must not project on one side of the wire more than on the other. It is not always easy to get a fragment satisfying these conditions, and at the same time suitable for mounting. A drop of solder occasionally assists the process of setting the diamond.

In drilling, the diamond must be held against the work by a constant force, applied either by means of weight or a spring. I made many trials by this method, using a watchmaker's lathe and pressing up the work by a weight and string, which passed over a pulley. I used about 40 ounces, and drilled a hole 3/32 in diameter in flint glass at a speed of 900 revolutions per minute to a depth of one-eighth of an inch in eight minutes. I used soap and water as a lubricant, and the work was satisfactory.

Since this was set up, I have been informed by Mr. Hicks of Hatton Garden that it is necessary to anneal glass rod by heating it up to the softening point and allowing it to cool very slowly under red-hot sand or asbestos before boring. If this be done, no trouble will be experienced. The annealing must be perfect.]

39. For boring large holes through thick glass sheets,

or, indeed, through anything where it is necessary to make sure that no accident can happen, or where great precision of position and form of hole is required, I find a boring tube mounted as shown in the picture (Fig. 36) is of great service. Brass or iron tube borers do perfectly well, and the end of the spindle may be provided once for all with a small tube chuck, or the tubes may be separately mounted as shown. A fairly high speed is desirable, and may be obtained either by foot, or, if power is available, is readily got by connecting to the speed cone of a lathe, which is presumably permanently belted to the motor.

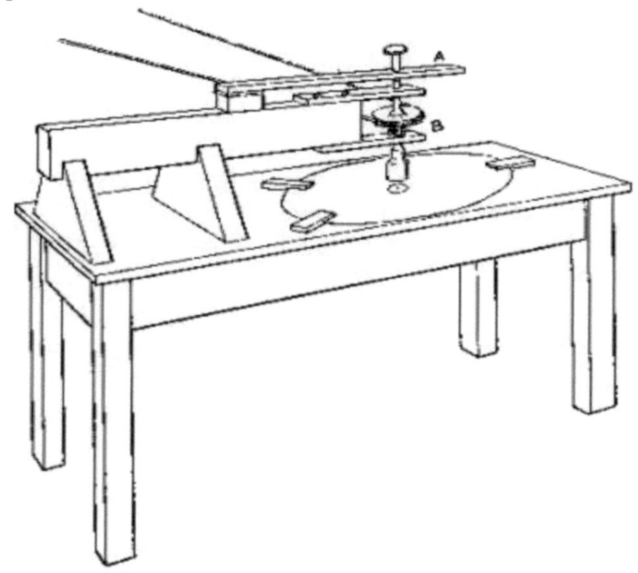

After trying tubes armed with diamond dust, as will be presently explained, I find that emery and thin oil or turpentine, if liberally supplied below the glass, will do very nearly as well. The tube should be allowed to rise from the work every few seconds, so as to allow of fresh emery and oil being carried into the circular grooves. This is done by lifting the hinged upper bearing, the drill being lifted by a spiral spring between the pulley and the lower bearing shown at B. The glass may be conveniently supported on a few sheets of paper if flat, or held firm in position by wooden clamps if of any other shape. In any case it should be firmly held down and should be well supported. Any desired pressure upon the drill is obtained by weighting the hinged board A.

40. The following method was shown to me by Mr. Wimshurst, but I have not had occasion to employ it myself. It is suitable for boring large holes through such glass as the plates of Mr. Wimshurst's Influence machines are usually made of. A diamond is mounted as the "pencil" of a compass, and with this a circle is drawn on the glass in the desired position. The other leg of the compass of course rests on a suitable washer.

To the best of my recollection the further procedure was as follows. A piece of steel rod about one-eighth inch in diameter was ground off flat and mounted in a vice vertically, so as to cause its plane end to form a small horizontal anvil. The centre (approximately) of the diamond-cut circle of the glass was laid on this anvil so as to rest evenly upon it, and the upper surface (i.e. that containing the cut) was then struck smartly with a hammer, completely pulverising the glass above the anvil. The hole was gradually extended in a similar manner right up to the diamond cut, from which, of course, the glass broke away.

A similar method has long been known to glaziers, differing from the preceding in that a series of diamond cuts are run across the circle parallel to two mutually perpendicular diameters. A smart tap on the back of the scored disc will generally cause the fragments to tumble out. I have never tried this myself, but I have seen it done.

Large discs may easily be cut from sheet glass by drawing a circular diamond cut, and gradually breaking away the outer parts by the aid of additional cuts and a pair of pliers or "shanks" (see Fig. 44).

41. Operations depending on Grinding: Ground-in Joints. —

The process will be perfectly understood by reference to a simple case. Suppose it is desired to grind the end of a tube into the neck of a bottle. If a stoppered bottle is available, the stopper must be taken out and measured as to its diameter at the top and bottoM. Select a bit of tube as nearly as possible of the same diameter as the stopper at its thickest part. Draw down the glass in the blow-pipe flame rather by allowing it to sink than by pulling it out. After a few trials no difficulty will be experienced in making its taper nearly equal to that of the stopper, though there will in all probability be several ridges and inequalities. When this stage is reached anneal the work carefully and see that the glass is not too thin. Afterwards use emery and water, and grind the stopper into the bottle.

There are six special directions to be note

(1)Turn the stopper through at least one revolution in each direction.

(2) Lift it out often so as to give the fresh emery a chance of getting into the joint.

(3) Rotate the bottle as well as the stopper in case there is any irregularity in the force brought to bear, which might cause one side of the neck to be more ground than another, or would cause the tube to set rather to one side or the other.

(4) Use emery passing a 50 sieve, i.e. a sieve with fifty threads to the inch run (see 144) to begin with, and when the stopper nearly fits, wash this thoroughly away, and finish with flour emery, previously washed to get rid of particles of excessive size; the process of washing will be fully discussed in the chapter on glass-grinding, which see.

(5) Any degree of fineness of surface may be obtained by using graded emery, as will be explained, but, in general, it is unnecessary to attempt a finer surface than can be got with washed flour emery. A superficial and imperfect polish may be given by grinding for a short time with powdered pumice stone.

(6) If the proper taper is not attained by blowing, or if ridges are left on the tapered part, the process may be both hastened and improved by giving the taper a preliminary filing with a three-cornered file and kerosene, just as one would proceed with iron or brass. A little filing will often save a good deal of grinding and make a better job.

If a bottle without a tapered neck is to be employed, it is as well to do the preliminary grinding by means of a cone turned up from a bit of cast iron. This is put in the lathe and pushed into the mouth of the bottle, the latter being supported by the hands. Use about the same surface speed as would be employed for turning cast iron. In this case the emery is better used with kerosene.

If a cylindrical bit of cast iron about an inch in diameter is turned down conically nearly to a point, it will save a good deal of trouble in making separate cones. If it gets ground into rings, and it becomes necessary to turn it up, use a diamond tool until the skin is thoroughly removed; the embedded emery merely grinds the edge off any ordinary steel tool.

For diamond tools see 55.

42. Use of the Lathe in Glass-working.

—

If it is necessary to remove a good deal of glass, time may be saved by actually turning the glass in a lathe. According to the direction given above for grinding a tube into the neck of a bottle, very little glass need be removed if the drawing down is well done, so that for this purpose turning is often unnecessary.

If the taper of the stopper be small and it is permissible to use a thick tube, or if a solid stopper only has to be provided, or an old stopper quickly altered to a new form, turning is very useful. The glass may be "chucked" in any suitable manner, and run at a speed not exceeding 10 feet per minute. Prepare a three-cornered file by mercury-hardening and by grinding the end flat so as to form a cutting angle of about 80, and use a moderate amount of kerosene lubrication, i.e. enough to keep the glass damp, but even this is not essential. Use the file as an ordinary brass turning tool, and press much more lightly than for metal turning. The glass will be found to scrape off quite pleasantly.

By chucking glass tubes on wooden mandrells the ends may be nicely turned in this manner ready for accurate closing by glass plates.

The process of grinding also is made much more rapid — at all events in the earlier stages — by chucking either the stopper or the bottle and holding the other member in the fingers, or in a wooden vice held in the hands. The finishing touches are best given by hand.

I ought to say that I think a good deal of glass-grinding, as practised in laboratories, might be advantageously replaced by glass turning or filing and certainly will be by any one who will give these methods a trial.

If one tube is to be ground into another, as in grinding a retort into a receiver, the latter must be drawn down from a larger piece, few beginners being able to widen a tube by the method explained with sufficient ease and certainty. The other operations are similar to the operations above described.

43. Funnels often require to be ground to an angle of 60. For this purpose it is well to keep a cast-iron cone, tapering from nothing up to four inches in diameter. This may be mounted on a lathe, and will be found of great use for grinding out the inside of funnels. Care must be taken to work the funnel backwards and forwards, or it will tend to grind so as to form rings, which interfere with filtering. A rough polish may be given on the lines explained in the next section.

44. A rough polish may be easily given to a surface which has been finished by washed flour emery, in the following manner. Turn up a disc of soft wood on the lathe, and run it at the highest wood-turning speed. Rub into the periphery a paste of sifted powdered pumice stone and water.

Any fairly smooth ground glass surface may be more or less polished by holding it for a moment against the revolving disc. Exact means of polishing will be described later on. Meanwhile this simple method will be found both quick and convenient, and is often quite sufficient where transparency, rather than figure, is required. I daresay a fine polish may be got on the same lines, using putty powder or washed rouge (not jewellers' rouge, which is too soft, but glass-polishers' rouge) to follow the pumice powder, but I have not required to try this.

45. It is sometimes required to give to ground glass surfaces a temporary transparency. This is to be done by using a film of oil of the same refractive index as the glass. Cornu has employed a varnish consisting of a mixture of turpentine and oil of cloves, but the yellow-brown colour of the latter is often a disadvantage. It will be found that a mixture of nut oil and oil of bitter almonds, or of bromo-napthalene and acetone, can be made of only a faint yellow colour; and by exact adjustment of the proportions will have the same refractive index for any ray as crown glass (ordinary window glass).

Procure a sample of the glass and smash it up to small fragments in an iron mortar. Sift out the fine dust and the larger pieces; bits about as large as small beads — say one-sixteenth inch every way — do very well. Boil the sifted glass with strong commercial hydrochloric acid to remove iron, wash with distilled water and a few drops of alcohol, dry on blotting paper in the sun or otherwise. Put the dry glass into a bottle or beaker, and begin by adding almond oil (or bromo-napthalene), then add nut oil (or acetone) till the glass practically disappears when examined by sodium light, or light of any other wave-length, as may be required.

The adjustment of the mixture is a matter of great delicacy, one drop too much of either constituent, in, say, 50 cubic centimetres, makes all the difference. The final adjustment is best accomplished by having two mixtures of the oils, one just too rich in almond, the other in nut oil; by adding one or other of these, the required mixture is soon obtained.

It is to be noted

(1) That adjustment is only perfect for light of one wave-length.

(2) That adjustment is only perfect at one temperature.

On examining a bottle of rather larger fragments of glass immersed in an adjusted mixture by ordinary daylight, a peculiarly beautiful play of colours is seen.

Of course, if it is only desired to make ground glass fairly transparent, these precautions are unnecessary, but it seemed better to dispose of the matter once for all in this connection.

M. Cornu's object was to make a varnish which would prevent reflection from the back of a photographic plate on to the film. I have had occasion to require to do the same when using a scale made by cutting lines through a film of black varnish on a slip of glass. This succeeded perfectly by making the varnish out of Canada balsam stained with a black aniline dye.

Mr. Russell, Government Astronomer of New South Wales, finds that the "halation" of star photographs can be prevented by pouring over the back of the plate a film of collodion suitably stained.

46. Making Ground Glass. —

This is easily done by rubbing the surface of polished glass with a bit of cast iron and washed "flour of emery." Of course, if the fineness of grain of the surface is of importance, appropriate sizes of emery must be employed. The iron may be replaced by a bit of glass cut with transverse grooves to allow the emery to distribute itself, or even by a bit of glass without such grooves, provided it does not measure more than one or two inches each way. If great speed is an object rather than the fineness of the surface, use a bit of lead and coarse emery, say any that will pass a sieve with fifty threads to the inch.

It may perhaps be mentioned here that it is a pity to throw away emery which has been used between glass and glass. In the chapter dealing with fine optical work the use of emery of various grades of fineness will be treated, and the finer grades can only be obtained (to my knowledge) from emery which has been crushed in the process of glass or metal grinding, especially the former. A large jam-pot covered with a cardboard lid does well as a receptacle of washings.

47. Glass-cutting. —

This is an art about which more can be learned in five minutes by watching it well practised than by pages of written description. My advice to any one about to commence the practice of the art would be to make friends with a glazier and see it done. What follows is therefore on the supposition that this advice has been followed.

After some experience of cutters made of especially hardened steel, I believe better work can generally be got out of a diamond, provided the cost is not an objection. It is economy to pay a good price for a good diamond. As is well known, the natural angle of the crystal makes the best point, and a person buying a diamond should examine the stone by the help of a lens, so as to see that this condition is fulfilled. The natural angle is generally, if not always, bounded by curved edges, which have a totally different appearance from the sharp edges of a "splinter."

When a purchase is to be made, it is as well for the student to take a bit of glass and a foot-rule with him, and to test the diamond before it is taken away. When a good diamond has been procured, begin by taking cuts on bits of clean window glass until the proper angle at which to hold the tool is ascertained. Never try to cut over a scratch, if you value your diamond, and never press hard on the glass; a good cut is accompanied by an unmistakable ringing sound quite different from the sound made when the diamond is only scratching.

Perhaps the most important advice that can be given is, *Never lend the diamond to anybody — under any circumstances.*

The free use of a diamond is an art which the physicist will do well to acquire, for quite a variety of apparatus may be made out of glass strips, and the accuracy with which the glass breaks along a good cut reduces such an operation as glass-box-making to a question of accurate drawing.

48. Cementing. —

One of the matters which is generally confused by too great a profusion of treatment is the art of cementing glass to other substances.

The following methods will be found to work, subject to two conditions:

(1) The glass must be clean;

(2) it must be hot enough to melt the cement.

For ordinary mending purposes when the glass does not require to be placed in water (especially if hot) nothing is better than that kind of glue which is generally called "diamond cement." This may be easily made by dissolving the best procurable isinglass in a mixture of 20 per cent water and 80 per cent glacial acetic acid — the exact proportions are not of consequence.

First, the isinglass is to be tightly packed into a bottle with a wide neck, then add the water, and let the isinglass soak it up. Afterwards pour in the acetic acid, and keep the mixture near 100C. for an hour or two on the water bath — or rather in it. The total volume of acetic acid and water should not be more than about half of the volume of isinglass when the latter is pressed into the bottle as tightly as possible.

The proper consistency of the cement may be ascertained by lifting a drop out of the bottle and allowing it to cool on a sheet of glass. In ten minutes it ought not to be more than slightly sticky, and the mass in the bottle, after standing a few hours cold, should not be sticky at all, and should yield, jelly-like, to the pressure of the finger to only a slight degree. If the glue is too weak, more isinglass may be added (without any preliminary soaking).

A person making the mixture for the first time almost always gets it too weak. It is difficult to give exact proportions by weight, as isinglass and gelatine (which may replace it) differ greatly in quality. This cement is applied like glue, and will cement nearly anything as well as glass. Of course, as much cement as possible must be squeezed out of any joint where it is employed. The addition of gums, as recommended in some books, is unnecessary.

Ordinary glue will serve perfectly for cementing glass to wood.

"Chipped glass" ware is, I understand, made by painting clean glass with glue. As the glue dries and breaks by contraction, it chips off the surface of the glass. I have never seen this done. In nearly all cases where alcohol is not to be employed very strong joints may be made by shellac. Orange shellac is stronger than the "bleached" variety.

A *sine qua non* is that the glass be hot enough to melt the shellac. The best way is to heat the glass surfaces and rub on the shellac from a bit of flake; the glass should not be so hot as to discolour the shellac appreciably, or its valuable properties will be partly destroyed. Both glass surfaces being thus prepared, and the shellac being quite fluid on both, they may be brought together and clamped tightly together till cool. Shellac that has been overheated, or dissolved in alcohol, or bleached, is of little use as compared with the pale orange flaky product. Dark flakes have probably been overheated during the preliminary refining.

For many purposes a cement is required capable of resisting carbon bisulphide. This is easily made by adding a little treacle (say 20 per cent) to ordinary glue. Since the mixture of glue and treacle does not keep, i.e. it cannot be satisfactorily melted up again after once it has set, no more should be made up than will be wanted at the time. If the glue be thick, glass boxes for carbon disulphide may be easily put together, even though the edges of the glass strips are not quite smooth, for, unlike most cements, this mixture remains tough, and is fairly strong in itself.

I have found by experiment that most fixed and, to a less degree, essential oils have little or no solvent action on shellac, and I suspect that the same remark applies to the treacle-glue mixture, but I have not tried. Turpenes act on shellac slightly, but mineral oils apparently not at all. The tests on which these statements are based were continued for about two years, during which time kerosene and mineral oils had no observable effect on shellac-fastened galvanometer mirrors.

49. Fusing Electrodes into Glass. —

This art has greatly improved since the introduction of the incandescent lamp; however, up to the present, platinum seems to remain the only substance capable of giving a certainly air-tight result. I have not tried the aluminium-alumina method.

Many years ago it was the fashion to surround the platinum wire with a drop of white enamel glass in order to cause better adhesion between it and the ordinary glass. *[Footnote:* Hittorf and Geissler (Pogg. Ann. 1864, 35; English translation, Phys. Soc. London, p. 138) found that it was impossible to make air-tight joints between platinum and hard potash glass, but that soft lead glass could be used with success as a cement.*]* However, in the case of flint glass, if one may judge from incandescent lamps, this is not essential — a fact which entirely coincides with my own experience.

On the other hand, when sealing electrodes into German glass I have often used a drop of enamel with perfect results, though this is not always done in Germany. In all cases, however, in which electrodes have to be sealed in — especially when they are liable to heat — I recommend flint glass, and in this have the support of Mr. Rain (*The Incandescent Lamp and its Manufacture*, p. 131). The exact details for the preparation of eudiometer tubes are given by Faraday (*Chemical Manipulation*, 1200).

In view of what has preceded, however, I will content myself with the following notes. Make the hole through which the wire is to protrude only slightly larger than the wire itself, and be sure that the latter is clean. Allow the glass to cool sufficiently not to stick to the wire when the latter is pushed in. Be sure that, on heating, the glass does not get reduced, and that it flows up to the wire all round; pull and push the wire a little with a pair of pincers, to ensure this.

It is not a bad plan to get the glass exceedingly fluid round the wire — even if the lump has to be blown out a little afterwards--as it cools. The seal should finally be well annealed in asbestos, but first by gradually moving it into the hot air in front of the flame.

It was observed by Professor J. J. Thomson and the author some years ago (*Proc*. Roy. Soc. 40. 331. 1886) that when very violent discharges are taken through lightly sealed-in electrodes in lead-glass tubes — say from a large battery of Leyden jars — gas appears to be carried into the tube over and above that naturally given off by the platinum, and this without there being any apparent want of perfection in the seal. This observation has since been confirmed by others. Consequently in experiments on violent discharges in vacuo where certainty is required as to the exclusion of air, the seals should be protected by a guard tube or cap containing mercury; this must, of course, be put in hot and clean, on hot and clean glass, and in special cases should be boiled in situ.

A well-known German physicist (Warburg, I think) recommends putting the seals under water, but I cannot think that this is a good plan, for if air can get in, why not water? which has its surface tension in its favour. The same reasoning prevents my recommending a layer of sulphuric acid above the mercury-a method used for securing air-tightness in "mercury joints" by Mr. Gimingham, Proc. R. S. 1874.

Further protection may be attained for many purposes by coating the platinum wire with a sheath of glass, say half an inch long, fused to the platinum wire to a depth of one-twentieth of an inch all round.

In some cases the electrodes must be expected to get very hot, for instance, when it is desired to platinise mirrors by the device of

Professor Wright of Yale. In this and similar cases I have met with great success by using "barometer" tubes of about one-twelfth of an inch bore, and with walls, say, one-tenth of an inch thick. *[Footnote: "Barometer" tube is merely very thick-walled glass tubing, and makes particularly bad barometers, which are sold as weather glasses.]*

This tube is drawn down to a long point — say an inch long by one-eighth of an inch external diameter, and the wire is fused in for a length, say, of three-quarters of an inch, but only in the narrow drawn-down part of the tube. At different times I have tried four such seals, and though the electrodes were red hot for hours, I have never had an accident — of course they were well annealed.

Fig.

37.

For directions as to the making of high vacuum tubes, see the section dealing with that matter.

50. As economy of platinum is often of importance, the following little art will save money and trouble. Platinum is easily caused to join most firmly to copper — with which, I presume, it alloys — by the following method. Hold the platinum wire against the copper wire, end to end, at the tip of the reducing flame of a typical blow-pipe — or anywhere — preferably in the "reducing" part of the ox-ygas flame; in a moment the metals will fuse together at the point of contact, when they may be withdrawn.

Such a joint is very strong and wholly satisfactory, much better than a soldered joint. If the work is not carried out successfully so that a considerable drop of copper-platinum alloy accumulates, cut it off and start again. The essence of success is speed, so that the copper does not get "burned." If any considerable quantity of alloy is formed it dissolves the copper, and weakens it, so that we have first the platinum wire, then a bead of alloy, and then a copper wire fused into the bead, but so thin just outside the latter that the joint has no mechanical strength.

51. The Art of making Air-light Joints.

—

Lamp-manufacturers and others have long since learned that when glass is in question not only are fused joints made as easily as others, but that they afford the only reliable form of joint. An experimenter who uses flint glass, has a little experience, an oxygas blow-pipe and a blowing apparatus, will prefer to make his joints in this way, simply from the ease with which it may be done. When it comes to making a tight joint between glass and other substances the problem is by no means so easy. Thus Mr. Griffiths (Phil. Trans. 1893, p. 380) failed to make air-tight joints by cementing glass into steel tubes, using hard shellac, and the tubes fitting closely. These joints were satisfactory at first, but did not last; the length of the joint is not stated. The difficulty was finally got over by soldering very narrow platinum tubes into the steel, and fusing the former into the glass.

Mr. Griffiths has since used an alloy with success as a cement, but I cannot discover what it is made from. Many years ago Professor Hittorf prepared good high vacuum tubes by plugging the ends of glass tubes with sealing wax merely, though in all cases the spaces to be filled with wax were long and narrow (Hittorf, Pogg. Ann. 1869, 5, English translation, Phys. Soc. p. 113). Again, Regnault habitually used brass ferules, and cemented glass into them by means of his mastic, which can still be procured at a low rate from his instrument-makers (Golan, Paris). Lenard also, in his investigations on Cathode Rays (Wied. Ann., vol. li. p. 224), made use of sealing wax covered with marine glue.

Surely in face of these facts we must admit that cement joints can be made with fair success. I do not know the composition of M. Regnault's mastic, but Faraday (Manipulations, 1123) gives the following receipt for a cement for joining ferules to retorts, etc. —

> Resin 5 parts
>
> Beeswax 1 part
>
> Red ochre or Venetian red,
> finely powdered and sifted 1 part

I believe this to be substantially the same as Regnault's mastic, though I have never analysed the latter.

For chemical work the possibility of evolution of gas from such a cement must be taken into account, and I should certainly not trust it for this reason in vacuum tube work, where the purity of the confined gas could come in question. Otherwise it is an excellent cement, and does not in my experience tend to crack away from glass to the same extent as paraffin or pure shellac.

This cracking away from glass, by the way, is probably an effect of difference in rate of expansion between the glass and cement which probably always exists, and, if the cement be not sufficiently viscous, must, beyond certain temperature limits, either produce cracks or cause separation. Professor Wright of Yale has used a hard mineral pitch as a cement in vacuum work with success.

My attention has been directed to a fusible metal cement containing mercury, and made according to the following receipt, given by Mr. S. G. Rawson, Journal of the Society of Chemical Industry, vol. ix. (1890), P. 150:—

> Bismuth 40 per cent
>
> Lead 25 per cent
>
> Tin 10 per cent
>
> Cadmium 10 per cent
>
> Mercury 15 per cent

This is practically one form of Rose's fusible metal with 15 per cent mercury added. It takes nearly an hour to set completely, and the apparatus must be clean and warm before it is applied.

As the result of several trials by myself and friends, I am afraid I must dissent from the claim of the author that such a cement will make a really air-tight joint between glass tubes. Indeed, the appearance of the surface as viewed through the glass is not such as to give any confidence, no matter what care may have been exercised in performing all the operations and cleaning the glass; besides which the cement is rigid when cold, and the expansion difficulty comes in.

On the other hand, if extreme air-tightness is not an object, the cement is strong and easily applied, and has many uses. I have an idea that if the joints were covered with a layer of soft wax, the result would be satisfactory in so far as air-tightness is concerned.

This anticipation has since been verified.

In many cases one can resort to the device already mentioned of enclosing a rubber or tape-wrapped joint between two tubes in a bath of mercury, but in this case the glass must be clean and hot and the mercury also warm, dry, and pure when the joint is put together, otherwise an appreciable air film is left against the glass, and this may creep into the joint.

Perhaps the easiest way of making such a joint is to use an outer tube of thin clean glass, and bore a narrow hole into it from one side to admit the mercury; if the mercury is to be heated in vacuo, it is better to seal on a side joint. It is always better, if possible, to boil the mercury in situ, which involves making the wrapping of asbestos, but, after all, we come back to the position I began by taking up, viz. that the easiest and most reliable method is by fusion of the glass — all the rest are unsuitable for work of real precision.

I should be ungrateful, however, were I not to devote a few lines to the great convenience and merit of so-called "centering cement." This substance has two or three very valuable properties. It is very tough and strong in itself, and it remains plastic on cooling for some time before it really sets. If for any reason a small tube has to be cemented into a larger one, which is a good deal larger, so that an appreciable mass of cement is necessary, and particularly if the joint requires to have great mechanical strength, this cement is invaluable. I have even used a plug of it instead of a cork for making the joint between a gas delivery tube and a calcium chloride tower. (Why are these affairs made with such abominable tubulures?)

The joint in question has never allowed the tube to sag though it projects horizontally to a distance of 6 inches, and has had to withstand nearly two years of Sydney temperature. The cement consists of a mixture of shellac and 10 per cent of oil of cassia.

The shellac is first melted in an iron ladle, and the oil of cassia quickly added and stirred in, to an extent of about 10 per cent, but the exact proportions are not of importance. Great care must be taken not to overheat the shellac.

APPENDIX TO CHAPTER I

ON THE PREPARATION OF VACUUM TUBES FOR THE PRODUCTION OF PROFESSOR ROENTGEN'S RADIATION

[Footnote: Written in May 1896.]

WHEN Professor Roentgen's discovery was first announced at the end of 1895 much difficulty was experienced in obtaining radiation of the requisite intensity for the repetition of his experiments. The following notes on the production of vacuum tubes of the required quality may therefore be of use to those who desire to prepare their own apparatus. It appears that flint glass is much more opaque to Roentgen's radiation than soda glass, and consequently the vacuum tubes require to be prepared from the latter material.

Fig.

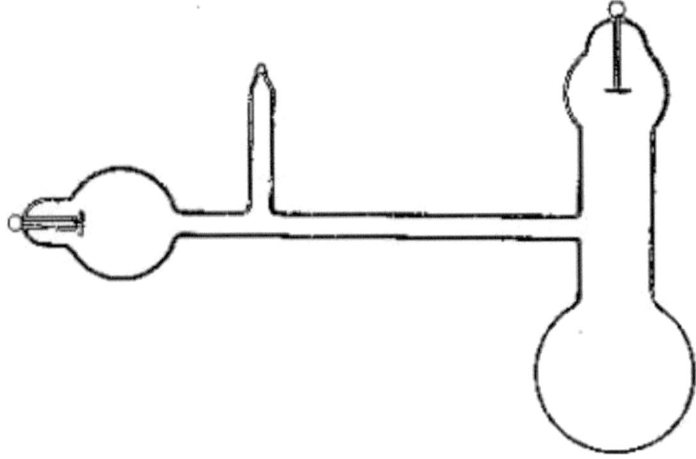

39.

A form of vacuum tube which has proved very successful in the author's hands is sketched in Fig. 38. It is most easily constructed as follows. A bit of tubing about 2 centimetres diameter, 15 centimetres long, and 1.5 millimetre wall thickness, is drawn down to a point. The larger bulb, about 5 centimetres in diameter, is blown at one end of this tube. The thinner the bulb the better, provided that it does not collapse under atmospheric pressure. A very good idea of a proper thickness may be obtained from the statement that about 4 centimetres length of the tubing should be blown out to form the bulb. This would give a bulb of about the thickness of an ordinary fractionating bulb. Before going any further it is as well to test the bulb by tapping on the table and by exhausting it by means of an ordinary water-velocity pump.

The side tube is next prepared out of narrower tubing, and is provided with a smaller bulb, a blowing-out tube, and a terminal, to be made as will be described. This side tube is next fused on to the main tube, special care being taken about the annealing, and the cathode terminal is then sealed into the main tube. After using clean glass it is in general only necessary to rinse the tube out with clean alcohol, after which it may be dried and exhausted.

The success of the operation will depend primarily on the attention given to the preparation and sealing-in of the electrode facing the large bulb.

Preparation of Terminals. — Some platinum wire of about No. 26 B.W.G. — the exact size is unimportant — must be provided, also some sheet aluminium about 1 millimetre thick, some white enamel cement glass, and a "cane" of flint-glass tube of a few millimetres bore.

The electrodes are prepared by cutting discs of aluminium of from 1 to 1.5 centimetres diameter. The discs of aluminium are bored in the centre, so as to admit the "stems" which are made of aluminium wire of about 1 millimetre diameter. The stems are then riveted into the discs. The "stems" are about I centimetre long, and are drilled to a depth of about 3 millimetres, the drill used being about double the diameter of the platinum wire to be used for making the connections. The faces of the electrodes — i.e. the free surfaces of the aluminium discs — are then hammered flat and brought to a burnished surface by being placed on a bit of highly polished steel and struck by a "set" provided with a hole to allow of the "stem" escaping damage. The operation will be obvious after a reference to Figs. 39 and 40; it is referred to again on page 96.

The platinum wires may be most conveniently attached by melting one end of the piece of platinum wire in the oxygas blow-pipe till it forms a bead just large enough to pass into the hole drilled up the stem of the electrode. The junction between the stein and the platinum wire is then made permanent by squeezing the aluminium down upon the platinum wire with the help of a pair of pliers. It is also possible to fuse the aluminium round the platinum, but as I have had several breakages of such joints, I prefer the mechanical connection described.

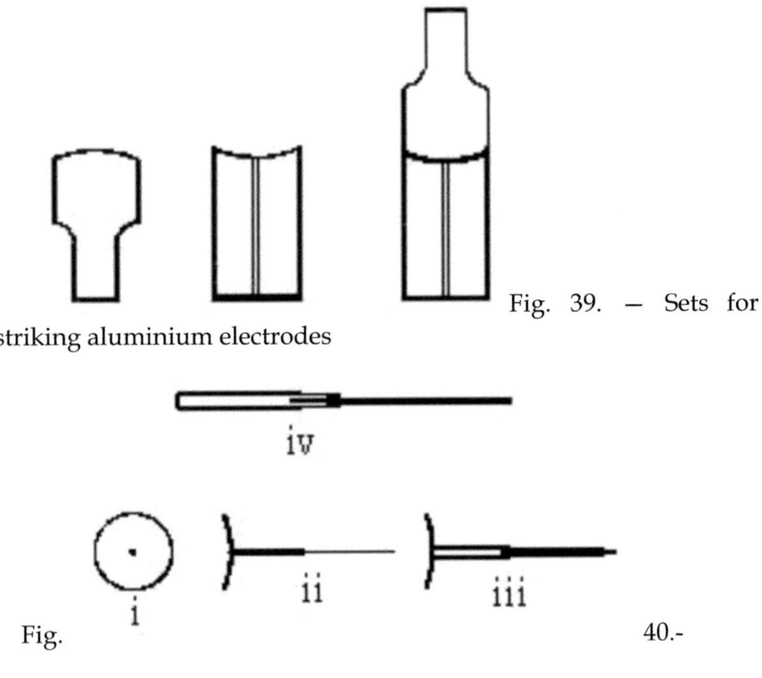

Fig. 39. — Sets for striking aluminium electrodes

Fig. 40.-

i.   Aluminium electrode.
ii.   Aluminium electrode connected to platinum wire.
iii.   Aluminium electrode connected to platinum wire and protected by glass.
iv.   Detail of fastening platinum wire.

The stem and platinum wire may now be protected by covering them with a little flint glass. For this purpose the flint-glass tube is pulled down till it will just slip over the stem and wire, and is cut off so as to leave about half a centimetre of platinum wire projecting. The flint-glass tube is then fused down upon the platinum wire, care being taken to avoid the presence of air bubbles. At the close of the operation a single drop of white enamel glass is fused round the platinum wire at a high temperature, so as to make a good joint with the protecting flint-glass tube.

The negative electrode being nearly as large as the main tube, it must be introduced before the latter is drawn down for sealing. After drawing down the main tube in the usual manner, taking care not to make it less than a millimetre in wall-thickness, it is cut off so as to leave a hole not quite big enough for the enamel drop to pass through. By heating and opening, the aperture is got just large enough to allow the enamel drop to pass into it, and when this is the case the joint is sealed, pulled, and blown out until the electrode occupies the right position — viz. in the centre of the tube and with its face normal to the axis of the tube.

The glass walls near the negative electrode must not be less than a millimetre thick, and may be rather more with advantage, the glass must be even, and the joint between the flint glass and the soda glass, or between the wire and the soda glass, must be wholly through the enamel. The "seal" must be well annealed. It will be found that the sealing-in process is much easier when the stem of the electrode is short and when the glass coating is not too heavy. Half a millimetre of glass thickness round the stein is quite sufficient.

The diagram, of the tube shows that the main tube has been expanded round the edges of the cathode. This is to reduce the heating consequent on the projection of cathode rays from the edges of the disc against the glass tube.

The anode is inserted into its bulb in a quite similar manner. If desired it may be made considerably smaller, and does not need the careful adjustment requisite in sealing-in the cathode, nor does the glass near the entry wire require to be so thick.

More intense effects are often got by making the cathode slightly concave, but in this case the risk of melting the thin glass is considerably increased. No doubt, Bohemian glass might be used throughout instead of soft soda glass, and this would not melt so easily; the difficulties of manipulating the glass are, however, more pronounced.

It will be shown directly that the best Roentgen effects are got with a high vacuum, and it is for this reason that the glass near the cathode seal requires to be strong. The potential right up to the cathode is strongly positive inside the tube, and this causes the glass to be exposed to a strong electric stress in the neighbourhood of the seal.

Although the GLASS-BLOWING involved in the making of a so-called focus tube is rather more difficult than in the case just described, there is no reason why such a difficulty should not be overcome; I will therefore explain how a focus tube may be made.

Fig.

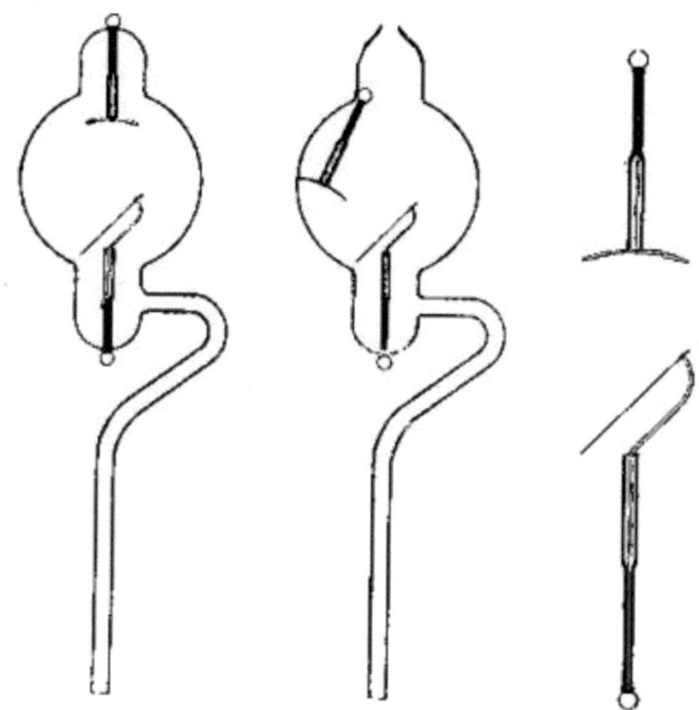

41.

A bulb about 3 inches in diameter is blown from a bit of tube of a little more than 1 inch diameter. Unless the walls of the tube are about one-eighth of an inch in thickness, this will involve a preliminary thickening up of the glass. This is not difficult if care be taken to avoid making the glass too hot. The larger gas jet described in connection with the soda-GLASS-BLOWING table must be employed. In blowing a bulb of this size it must not be forgotten that draughts exercise a very injurious influence by causing the glass to cool unequally; this leads to bulbs of irregular shape.

In the method of construction shown in Fig. 41, the anode is put in first. This anode simply consists of a square bit of platinum or platinum-iridium foil, measuring about 0.75 inch by 1 inch, and riveted on to a bent aluminium wire stem.

As soon as the anode is fused in, and while the glass is still hot, the side tube is put on. The whole of the anode end is then carefully annealed. When the annealing is finished the side tube is bent as shown to serve as a handle when the time comes to mount the cathode. Before placing the cathode in position, and while the main tube is still wide open, the anode is adjusted by means of a tool thrust in through this open end. This is necessary in view of the fact that the platinum foil is occasionally bent during the operation of forcing the anode into the bulb.

The cathode is a portion of a spherical surface of polished aluminium, a mode of preparing which will be given directly. The cathode having been placed inside the bulb, the wide glass tube is carefully drawn down and cut off at such a point that when the cathode is in position its centre of curvature will lie slightly in front of the anode plate. For instance, if the radius of curvature of the cathode be 1.5 inches, the centre of curvature may lie something like an eighth of an inch or less in front of the anode.

The cathode as shown in Fig. 41 is rather smaller than is advantageous. To make it much larger than is shown, however, the opening into the bulb would require to be considerably widened, and though this is not really a difficult operation, still it requires more practice than my readers are likely to have had. The difficulty is not so much in widening out the entry as in closing it down again neatly.

Now as to making the anode. A disc of aluminium is cut from a sheet which must not be too thick — one twenty-fifth of an inch is quite thick enough. This disc is bored at the centre to allow of the stem being riveted in position. The disc is then annealed in the Bunsen flame and the stem riveted on.

The curvature is best got by striking between steel dies (see Figs. 39 and 40). Two bits of tool steel are softened and turned on the lathe, one convex and the other concave. The concave die has a small hole drilled up the centre to admit the stem. The desired radius of curvature is easily attained by cutting out templates from sheet zinc and using them to gauge the turning. The two dies are slightly ground together on the lathe with emery and oil and are then polished, or rather the convex die is polished — the other one does not matter. The polishing is most easily done by using graded emery and oil and polishing with a rag. The method of grading emery will be described in the chapter on glass-grinding.

The aluminium disc is now struck between the dies by means of a hammer. If the radius of curvature is anything more than one inch and the disc not more than one inch in diameter the cathode can be struck at once from the flat as described. For very deep curves no doubt it will be better to make an intermediate pair of dies and to re-anneal the aluminium after the first striking.

When the tube is successfully prepared so far as the glassblowing goes it may be rinsed with strong pure alcohol both inside and out, and dried. The straight part of the side tube is then constricted ready for fusing off and the whole affair is placed on the vacuum pump.

In spite of the great improvements made during recent years in the construction of so-called Geissler vacuum pumps — i.e. pumps in which a Torricellian vacuum is continually reproduced — I am of opinion that Sprengel pumps are, on the whole, more convenient for exhausting Crooke's tubes. A full discussion of the subject of vacuum pumps will be found in a work by Mr. G. S. Ram (The Incandescent Lamp and its Manufacture), published by the Electrician Publishing Company, and it is not my intention to deal with the matter here; the simplest kind of Sprengel pump will be found quite adequate for our purpose, provided that it is well made.

Fig. 42 is intended to represent a modification of a pump based on the model manufactured by Hicks of Hatton Garden, and arranged to suit the amateur glass-blower. The only point of importance is the construction of the head of the fall tube, of which a separate and enlarged diagram is given. The fall tubes may have an internal diameter up to 2 mm. (two millimetres) and an effective length of 120 cm.

Free use is made of rubber tube connections in the part of the pump exposed to the passage of mercury. The rubber employed should be black and of the highest quality, having the walls strengthened by a layer of canvas. If such tube cannot be easily obtained, a very good substitute may be made by placing a bit of ordinary black tube inside another and rather larger bit and binding the outer tube with tape or ribbon. In any case the tubing which comes in contact with the mercury should be boiled in strong caustic potash or soda solution for at least ten minutes to get rid of free sulphur, which fouls the mercury directly it comes in contact with it. The tubing is well washed, rinsed with alcohol, and carefully dried.

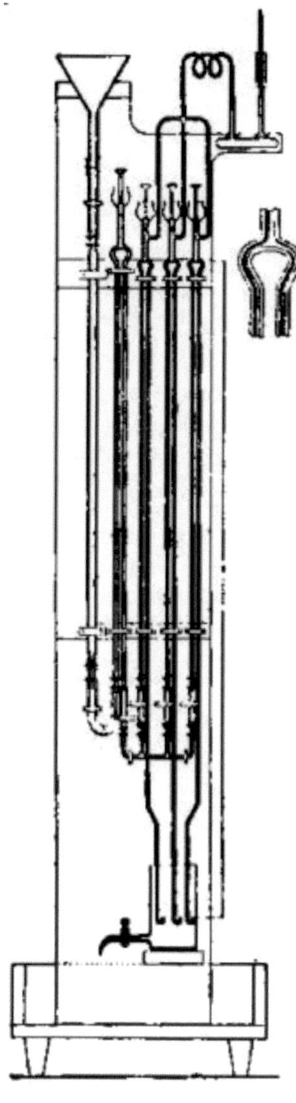

Fig. 42.

The diagram represents what is practically a system of three Sprengel pumps, though they are all fed from the same mercury reservoir and run down into the same mercury receiver. It is much easier to make three pumps, each with separate pinch cocks to regulate the mercury supply, than it is to make three jets, each delivering exactly the proper stream of mercury to three fall tubes.

Sprengel pumps only work at their highest efficiency when the mercury supply is carefully regulated to suit the peculiarities of each fall tube, and this is quite easily done in the model figured. Since on starting the pump the rubber connections have to stand a considerable pressure, the ends of the tubes must be somewhat corrugated to enable the rubber to be firmly wired on to them. The best binding wire is the purest Swedish iron wire, previously annealed in a Bunsen gas flame.

The wire must never be twisted down on the bare rubber, but must always be separated from it by a tape binding. By taking this precaution the wire maybe twisted very much more tightly than is otherwise possible without cutting the rubber.

The only difficulty in making such a pump as is described lies in the bending of the heads of the fall tubes. This bending must be done with perfect regularity and neatness, otherwise the drops of mercury will not break regularly, or will break just inside the top of the fall tube, and so obstruct its entrance that at high vacua no air can get into the tube at all.

The connections at the head of the fall tubes must also be well put on and the joints blown out so that the mercury in dropping over the head is not interfered with by the upper surface of the tube. However, a glance at the enlarged diagram will show what is to be aimed at better than any amount of description. In preparing the fall tubes it is generally necessary to join at least two "canes" together. The joint must be arranged to occur either in the tube leading the mercury to the head of the fall, or in that part of the fall tube which remains full of mercury when the highest vacuum is attained. On no account must the joint be made at the fall itself (at least not by an amateur), nor in that part of the fall tube where the mercury falls freely, particularly at its lower end, where the drops fall on the head of the column of mercury.

When a high vacuum is attained the efficacy of the pump depends chiefly on the way in which the drops fall on the head of the column. If the fall is too long the drops are apt to break up and allow the small bubble of air to escape up the tube, also any irregularity or dirt in the tube at this point makes it more easy for the bubbles of air to escape to the surface of the mercury.

Any pump in which the supply of mercury to the fall tube can be regulated nicely will pump well until the lowest available pressures are being attained; a good pump will then continue to hold the air bubbles, while a bad one will allow them to slip back *[Footnote: For special methods of avoiding this difficulty see Mr. Ram's book.]* ...

Though three fall tubes are recommended, it must not be supposed that the pump will produce a Crooke's vacuum three times more rapidly than one fall tube. Until the mercury commences to hammer in the pump the three tubes will pump approximately three times faster than one tube, but as soon as the major portion of the air collected begins to come from the layer condensed on the glass surface of the tube to be exhausted and from the electrodes, the rate at which exhaustion will go on no longer depends entirely on the pump.

In order that bubbles of air may not slip back up the fall tube it is generally desirable to allow the mercury to fall pretty briskly, and in this case the capacity of the pump to take air is generally far in excess of the air supply. One advantage of having more than one fall tube is that it often happens that a fall tube gets soiled during the process of exhaustion and no longer works up to its best performance. Out of three fall tubes, however, one is pretty sure to be working well, and as soon as the mercury begins to hammer in the tubes the supply may be shut off from the two falls which are working least satisfactorily.

Thus we are enabled to pump rapidly till a high degree of exhaustion is attained, having practically three pumps instead of one, whereas when the final stages are reached, and three pumps are only a drawback in that they increase the mercury flow, the apparatus is capable of instant modification to meet the new conditions.

The thistle funnels at the head of the fall tubes are made simply by blowing bulbs and then blowing the heads of the bulbs into wider ones, and finally blowing the heads of the wider bulbs off by vigorous blowing. The stoppers are ground in on the lathe before the tubes are attached to the fall tubes. The stoppers require to be at least half an inch long where they fit the necks, and must be really well ground in. The stoppers must first be turned up nicely and the necks ground out by a copper or iron cone and emery. The stoppers are rotated on a lathe at quite a slow speed, say 30 or 40 feet per minute, and the necks are held against them, as described in the section dealing with this art. The stoppers must in this case be finished with "two seconds" emery, and lastly with pumice dust and water (see chapter on glass-grinding).

Unless the stoppers fit exceedingly well trouble will arise from the mercury (which is poured into the thistle heads to form a seal) being forced downwards into the pump by atmospheric pressure.

The joints between the three fall tubes and the single exhaust main are easily made when the tubes are finally mounted, the hooked nozzle of the oxygas blow-pipe being expressly made for such work.

It is, on the whole, advisable to make the pump of flint glass, or at all events the air-trap tube and the fall tubes. A brush flame from the larger gas tube of the single blowpipe table is most suitable for the work of bending the tubes. The jointing of the long, narrow bore fall tubes is best accomplished by the oxygas flame, for in this way the minimum of irregularity is produced; the blowing tubes will of course be required for the job, and the narrow tubes must be well cleaned to begin with.

The air trap is an important though simple part of the pump. Its shoulder or fall should stand rather higher than the shoulders of the fall tubes, so that the mercury may run in a thin stream through a good Torricellian vacuum before it passes down to the fall tubes. This is easily attained by regulating the main mercury supply at the pinch cock situated between the tube from the upper reservoir and the air-trap tube, the other cocks being almost wide open.

It might be thought that the mercury would tend to pick up air in passing through the rubber connections to the fall tubes, but I have not found this to be the case in practice. There is, of course, no difficulty in eliminating the rubber connections between the fall tubes and the mercury supply from the air trap, but it impresses a greater rigidity on the structure and, as I say, is not in general necessary. It must not be forgotten that the mercury always exercises considerable pressure on the rubber joints, and so there is little tendency for gas to come out of the rubber.

The thistle funnels at the head of the fall tubes provide a simple and excellent means of cleaning the fall tubes. For this purpose some "pure" sulphuric acid which has been boiled with pure ammonium sulphate is placed in each thistle funnel, and when the fall tube is dirty the connection to the mercury supply is cut off at the pinch cock so as to leave the tube between this entry and the head of the fall tube quite full of mercury, and the sulphuric acid is allowed to run down the fall tube by raising the stopper. The fall tube should be allowed to stand full of acid for an hour or so, after which it will be found to be fairly clean.

Of course the mercury reservoir thus obtains a layer of acid above the mercury, and as it is better not to run the risk of any acid getting into the pump except in the fall tubes, the reservoir is best emptied from the bottom, by a syphon, if a suitable vessel cannot be procured, so that clean mercury only is withdrawn.

The phosphorus pentoxide tube is best made as shown simply from a bit of wide tube, with two side connections fused to the rest of the pump. It is no more trouble to cut the tube and fuse it up again when the drying material is renewed than to adjust the drying tube to two fixed stoppers, which is the alternative. The practice here recommended is rendered possible only by the oxygas blowpipe with hooked nozzle. The connection between the pump and tube to be exhausted is made simply by a short bit of rubber tube immersed in mercury.

The phosphorus pentoxide should be pure, or rather free from phosphorus and lower oxides; unless this be the case, the vapour arising from it is apt to soil the mercury in the pump. The phosphorus pentoxide is purified by distilling with oxygen over red-hot platinum black; if this cannot be done, the pentoxide should at least be strongly heated in a tube, in a current of dry air or oxygen, before it is placed in the drying tube.

The mercury used for the pump must be scrupulously clean. It does not, however, require to have been distilled in vacuo. It is sufficient to purify it by allowing it to fall in a fine spray into a large or rather tall jar of 25 per cent nitric acid and 75 per cent water. The mercury is then to be washed and dried by heating to, say, 110 C. in a porcelain dish.

Exhausting a Roentgen Tube. —

With a pump such as has been described there is seldom any advantage in fusing an extra connection to the vacuum tube so as to allow of a preliminary exhaustion by means of a water pump. About half an hour's pumping may possibly be saved by making use of a water pump.

The tube to be exhausted is washed and dried by careful heating over a Bunsen burner and by the passage of a current of air. The exhausting tube is then drawn down preparatory to sealing off, and the apparatus placed upon the pump. It is best held in position by a wooden clamp supported by a long retort stand.

Exhaustion may proceed till the mercury in the fall tubes commences to hammer. At this point the tube must be carefully heated by a Bunsen flame, the temperature being brought up to, say, 400 C. The heating may be continued intermittently till little or no effect due to the heating is discernible at the pump. When this stage is reached, or even before, the electrodes may be connected up to the coil and a discharge sent through the tube.

Care must be taken to stop the discharge as soon as a purple glow begins to appear, because when this happens, the resistance of the tube is very low, the electrodes get very hot, and may easily get damaged by a powerful discharge, and the platinum of the anode (if a focus tube is in question) begins to be distilled on to the glass. The heating and sparking are to be continued till the resistance of the tube sharply increases. This is tested by always having a spark gap, conveniently formed by the coil terminals, in parallel with the tube. If the terminals are points, it is convenient to set them at about one quarter of an inch distance apart.

As soon as sparks begin to pass between the terminals of the spark gap it becomes necessary to watch the process of exhaustion very carefully. In the first place, stop the pump, but let the coil run, and note whether the sparks continue to flow over the terminals. If the glass and electrodes are getting gas free, the discharge will continue to pass by the spark gap, but if gas is still being freely given off, then in perhaps three minutes the discharge will return to the tube, and pumping must be recommenced. The Roentgen effect only begins to appear when the tube has got to so high a state of exhaustion that the resistance increases rapidly.

By pumping and sparking, the resistance of the tube may be gradually raised till the spark would rather jump over 2 inches of air than go through the tube. When this state is attained the Roentgen effect as tested by a screen of calcium tungstate should be very brilliant. No conclusion as to the equivalent resistance of the tube can be arrived at so long as the discharge is kept going continually. When the spark would rather go over an inch of air in the spark gap than through the tube the pumping and sparking may be interrupted and the tube allowed to rest for, say, five minutes. It will generally be found that the equivalent resistance of the tube will be largely increased by this period of quiescence. It may even be found that the spark will now prefer to pass an air gap 3 inches long.

In any case the sparking should now be continued, the pump being at rest, and the variations of tube resistance watched by adjusting the spark gap. If the resistance falls below an equivalent of 2 inches of air in the gap the pump must be brought into action again and continued until the resistance as thus estimated remains fairly constant for, say, ten minutes. When this occurs the narrow neck of the exhaust tube may be strongly heated till the blow-pipe flame begins to show traces of sodium light. The flame must then be withdrawn and the discharge again tested. This is necessary because it occasionally happens that gas is given off during the heating of the neck to the neighbourhood of its fusion temperature.

If all is right the neck may now be fused entirely off and the tube is finished. Tubes of the focus pattern with large platinum anodes are in general (in my experience) much more difficult to exhaust than tubes of the kind first described. This is possibly to be attributed mainly to the gas given off by the platinum, but is also, no doubt, due to the tubes being much larger and exposing a larger glass surface. The type of tube described first generally takes about two hours to exhaust by a pump made as explained, while a "focus" tube has taken as long as nine hours, eight of which have been consumed after the tube was exhausted to the hammering point.

The pressure at which the maximum heating of the anode by the cathode rays occurs is a good deal higher than that at which the maximum Roentgen effect is produced. There is little doubt that the Roentgen radiation changes in nature to some extent as the vacuum improves either as a primary or secondary effect. It is therefore of some importance to test the tube for the purpose for which it is to be used during the actual exhaustion. It has been stated, for instance, that the relative penetrability of bone and flesh to Roentgen radiation attains a maximum difference at a certain pressure; this is very likely the case. Whether this effect is a direct function of the density of the gas in the tube, or whether it is dependent on the voltage or time integral of the current during the discharge, are questions which still await a solution.

The preparation of calcium tungstate for fluorescent screens is very simple.

Commercial sodium tungstate is fused with dried calcium chloride in the proportion of three parts of the former to two parts of the latter, both constituents being in fine powder and well mixed together. The fusion is conducted in a Fletcher's crucible furnace in a clay crucible. The temperature is raised as rapidly as possible to the highest point which the furnace will attain — i.e. a pure white heat. At this temperature the mixture of salts becomes partly fluid, or at least pasty, and the temperature may be kept at its highest point for, say, a quarter of an hour. At the end of this time the mass is poured and scraped on to a brick, and when cold is broken up and boiled with a large excess of water to dissolve out all soluble matter. The insoluble part, which consists of a gray shining powder, is washed several times with hot water, and is finally dried on filter paper in a water oven.

In order to prepare a screen the powder is ground slightly with very dilute shellac varnish, and is then floated over a glass plate so as to get an even covering. Unless the covering be very even the screen is useless, and no pains should be spared to secure evenness. It is not exactly easy to get a regular coat of the fluorescent material, but it may be done with a little care.

CHAPTER II

GLASS-GRINDING AND OPTICIANS' WORK

52. As no instructions of any practical value in this art have, so far as I know, appeared in any book in English, though a great deal of valuable information has been given in the *English Mechanic* and elsewhere, I shall deal with the matter sufficiently fully for all practical purposes. On the other hand, I do not propose to treat of all the methods which have been proposed, but only those requisite for the production of the results claimed. The student is requested to read through the chapter before commencing any particular operation.

53. The simplest way will be to describe the process of manufacture of some standard optical appliance, from which a general idea of the nature of the operations will be obtained. After this preliminary account special methods may be considered in detail. I will begin with an account of the construction of an achromatic object glass for a telescope, not because a student in a physical laboratory will often require to make one, but because it illustrates the usual processes very well; and requires to be well and accurately made.

A knowledge of the ordinary principles of optics on the part of the reader is assumed, for there are plenty of books on the theory of lenses, and, in any case, it is my intention to treat of the art rather than of the science of the subject. By far the best short statement of the principles involved which I have seen is Lord Rayleigh's article on Optics in the Encyclopaedia Britannica, and this is amply sufficient.

The first question that crops up is, of course, the subject of the choice of glass. It is obvious that the glass must be uniform in refractive index throughout, and that it must be free from air bubbles or bits of opaque matter. [*Footnote:* The complete testing of glass for uniformity of refractive index can only be arrived at by grinding and polishing a sufficient portion of the surfaces to enable an examination to be made of every part. In the case of a small disc it is sufficient to polish two or three facets on the edge, and to examine the glass in a field of uniform illumination through the windows thus formed. Very slight irregularities will cause a "mirage" easily recognised.]

The simplest procedure is to obtain glass of the desired quality from Messrs. Chance of Birmingham, according to the following abbreviated list of names and refractive indices, which may be relied upon:—

	Density.	Refractive Index.			
		C	D	F	G
Hard crown	2.85	1.5146	1.5172	1.5232	1.5280
Soft crown	2.55	1.5119	1.5146	1.5210	1.5263
Light flint	3.21	1.5700	1.5740	1.5839	1.5922
Dense flint	3.66	1.6175	1.6224	1.6348	1.6453
Extra dense flint	3.85	1.6450	1.6504	1.6643	1.6761
Double extra dense flint	4.45	1.7036	1.7103	1.7273	...

The above glasses may be had in sheets from 0.25 to 1 inch thick, and 6 to 12 inches square, at a cost of, say, 7s. 6d. per pound.

Discs can also be obtained of any reasonable size. Discs 2 inches in diameter cost about 1 per dozen, discs 3 inches in diameter about 10s. each. The price of discs increases enormously with the size. A 16-inch disc will cost about 100.

For special purposes, where the desired quality of glass does not appear on the list, an application may be made to the Jena Factory of Herr Schott. In order to give a definite example, I may mention that for ordinary telescopic objectives good results may be obtained by combining the hard crown and dense flint of Chance's list, using the crown to form a double convex, and the flint to form a double concave lens. The convex lens is placed in the more outward position in the telescope, i.e. the light passes first through it.

The conditions to be fulfilled are:

1. The glass must be achromatic;
2. it must have a small spherical aberration for rays converging to the principal focus.

It is impossible to discuss these matters without going into a complete optical discussion. The radii of curvature of the surfaces, beginning with the first, i.e. the external face of the convex lens, are in the ratio of 1, 2, and 3; an allowance of 15 inches focal length per inch of aperture is reasonable (see Optics in Ency. Brit.), and the focal length is the same as the greatest radius of curvature. Thus, for an object glass 2 inches in diameter, the first surface of the convex lens would have a radius of curvature of 10 inches, the surface common to the convex and concave lens would have a radius of curvature of 20 inches, and the last surface a radius of curvature of 30 inches. This would also be about the focal length of the finished lens. The surfaces in contact have, of course, a common curvature, and need not be cemented together unless a slight loss of light is inadmissible.

I will assume that a lens of about 2 inches diameter is to be made by hand, i.e. without the help of a special grinding or polishing machine; this can be accomplished perfectly well, so long as the diameter of the glass is not above about 6 inches, after which the labour is rather too severe. The two glass discs having been obtained from the makers, it will be found that they are slightly larger in diameter than the quoted size, something having been left for the waste of working.

It is difficult to deal with the processes of lens manufacture without entering at every stage into rather tedious details, and, what is worse, without interrupting the main account for the purpose of describing subsidiary instruments or processes. In order that the reader may have some guide in threading the maze, it is necessary that he should commence with a clear idea of the broad principles of construction which are to be carried out. For this purpose it seems desirable to begin by roughly indicating the various steps which are to be taken.

(1) The glass is to be made circular in form and of a given diameter.

(2) *Called Rough Grinding.* — The surfaces of the glass are to be made roughly convex, plane, or concave, as may be required; the glass is to be equally thick all round the edge. In this process the glass is abraded by the use of sand or emery rubbed over it by properly shaped pieces of iron or lead called "tools."

(3) The glass is ground with emery to the correct spherical figure as given by a spherometer.

(4) *Called Fine Grinding.* — The state of the surface is gradually improved by grinding with finer and finer grades of emery.

(5) The glass is polished by rouge.

(6) The glass is "figured." This means that it is gradually altered in form by a polishing tool till it gives the best results as found by trial.

In processes 2 to 5 counterpart tool surfaces are required — as a rule two convex and two concave surfaces for each lens surface. These subsidiary surfaces are worked (i.e. ground) on discs of cast iron faced with glass, or on slate discs; and discs thus prepared are called "tools."

Taking these processes in the order named, the mode of manufacture is shortly as follows: —

(1) The disc of glass, obtained in a roughly circular form, is mounted on an ordinary lathe, being conveniently cemented by Regnault's mastic to a small face plate. The lathe is rotated slowly, and the glass is gradually turned down to a circular figure by means (1) of a tool with a diamond point; or (2) an ordinary hand-file moistened with kerosene, as described in 42; or (3) a mass of brass or iron served with a mixture of emery — or sand — and water fed on to the disc, so that the disc is gradually ground circular.

The operation of making a circular disc of given diameter does not differ in any important particular from the similar operation in the case of brass or iron, and is in fact merely a matter of turning at a slow speed.

(2 and 3) Roughing or bringing the surfaces of the glass roughly to the proper convex or concave shape. — This is accomplished by grinding, generally with sand in large works, or with emery in the laboratory, where the time saved is of more importance than the value of the emery.

Discs of iron or brass are cast and turned so as to have a diameter slightly less than that of the glass to be ground, and are, say, half an inch thick. These discs are turned convex or concave on one face according as they are to be employed in the production of concave or convex glass surfaces. The proper degree of convexity or concavity may be approximated to by turning with ordinary turning tools, using a circular arc cut from zinc or glass (as will be described) as a "template" or pattern. This also is a mere matter of turning.

The first approximation to the desired convex or concave surface of the glass is attained (in the case of small lenses, say up to three inches diameter) by rotating the glass on the lathe as described above (for the purpose of giving it a circular edge) and holding the tool against the rotating glass, a plentiful supply of coarse emery and water, or sand and water, being supplied between the glass and metal surfaces. The tool is held by hand against the surface of the revolving glass, and is constantly moved about, both round its own axis of figure and to and fro across the glass surface. In this way the glass gradually gets convex or concave.

The curvature is tested from time to time by a spherometer, and the tool is increased or decreased in curvature by turning it on a lathe so as to cause it to grind the glass more at the edges or in the middle according to the indications of the spherometer.

This instrument, by the way — so important for lens makers — consists essentially of a kind of three-legged stool, with an additional leg placed at the centre of the circle circumscribing the other three. This central leg is in reality a fine screw with a very large head graduated on the edge, so that it is easy to compute the fractions of a turn given to the screw. The instrument is first placed on a flat plate, and the central screw turned till its end just touches the plate, a state of affairs which is very sharply discernible by the slight rocking which it enables the instrument to undergo when pushed by the hand. See the sketch.

On a convex or concave surface the screw has to be screwed in or out, and from the amount of screwing necessary to bring all four points into equal contact, the curvature may be ascertained.

Let a be the distance between the equidistant feet, and d the distance through which the screw is protruded or retracted from its zero position on a flat surface. Then the radius of curvature ρ is given by the formula $2\rho = a^2/3d + d$.

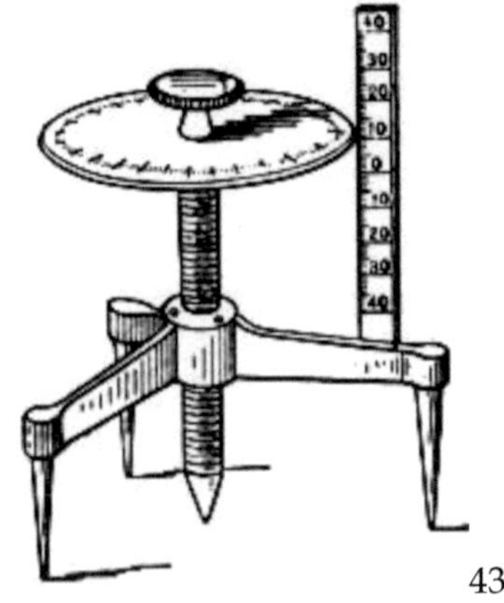

Fig. 43.

The process of roughing is not always carried out exactly as described, and will be referred to again.

(4) The glass being approximately of the proper radius of curvature on one side, it is reversed on the chuck and the same process gone through on the other side. After this the glass is usually dismounted from the lathe and mounted by cement on a pedestal, which is merely a wooden stand with a heavy foot, so that the glass may be held conveniently for the workman. Sometimes a pedestal about four feet high is fixed in the floor of the room, so that the workman engaged in grinding the lens may walk round and round it to secure uniformity. For ordinary purposes, however, a short pedestal may be placed on a table and rotated from time to time by hand, the operator sitting down to his work.

Rough iron or brass tools do not succeed for fine grinding — i.e. grinding with fine emery, because particles of emery become embedded in the metal so tightly that they cannot be got out by any ordinary cleaning. If we have been using emery passing say a sieve with 60 threads to the inch, and then go on to some passing say 100 threads to the inch, a few of the coarser particles will adhere to the "tool", and go on cutting and scratching all the time grinding by means of the finer emery is in progress.

To get over this it is usual to use a rather different kind of grinding tool. A very good kind is made by cementing small squares of glass (say up to half an inch on the side), on to a disc of slate slightly smaller than the lens surface to be formed (Fig. 51). The glass-slate tool is then "roughed" just like the lens surface, but, of course, if the lens has been roughed "convex" the tool must be roughed "concave".

The "roughed" tool is then used to gradually improve the fineness of grinding of the glass. For this purpose grinding by hand is resorted to, the tool and lens being supplied continually with finer and finer emery. Fig. 52 gives an idea of the way in which the tool is moved across the glass surface. Very little pressure is required. The tool is carried in small circular sweeps round and round the lens, so that the centre of the tool describes a many-looped curve on the lens surface. The tool must be allowed to rotate about its own axis; and the lens and pedestal must also be rotated from time to time.

Every few minutes the circular strokes are interrupted, and simple, straight, transverse strokes taken. In no case (except to correct a, defect, as will be explained) should the tool overhang the lens surface by more than about one quarter the diameter of the latter. After grinding say for an hour with one size of emery fed in by means of a clean stick say every five minutes, the emery is washed off, and everything carefully cleaned. The process is then repeated with finer emery, and so on.

The different grades of emery are prepared by taking advantage of the fact that the smaller the particles the longer do they remain suspended in water. Some emery mud from a "roughing" operation is stirred up with plenty of water and left a few seconds to settle, the liquor is then decanted to a second jug and left say for double the time, say ten seconds; it is decanted again, and so on till four or five grades of emery have been accumulated, each jug containing finer emery than its predecessor in the process.

It is not much use using emery which takes more than half an hour to settle in an ordinary bedroom jug. What remains in the liquid to be decanted is mostly glass mud and not emery at all. The process of fine grinding is continually checked by the spherometer, and the art consists in knowing how to move the grinding tool so as to make the lens surface more or less curved. In general it may be said that if the tool is moved in small sweeps, and not allowed to overhang much, the Centre of the lens will be more abraded, while if bold free strokes are taken with much overhanging, the edges of the lens will be more ground away.

By the exercise of patience and perseverance any one will succeed in gradually fine grinding the lens surface and keeping it to the spherometer, but the skill comes in doing this rapidly by varying the shape of the strokes before any appreciable alteration of curvature has come about.

Polishing. —

The most simple way of polishing is to coat the grinding tool with paper, as will be described, and then to brush some rouge into the paper. The polisher is moved over the work in much the same way as the fine grinding tool, until the glass is polished. Many operators prefer to use a tool made by squeezing a disc of slate, armed with squares of warm pitch, against the lens surface (finely ground), and then covering these squares with rouge and water instead of emery and water as in the fine grinding process.

The final process is called "figuring." It will in general be unnecessary with a small lens. With large lenses or mirrors the final touches have to be given after the optical behaviour of the lens or mirror has been tested with the telescope itself, and this process is called "figuring." A book might easily be written on the optical indications of various imperfections in a mirror or lens. Suffice it to say here that a sufficiently skilled person will be able to decide from an observation of the behaviour of a telescope whether a lens will be improved by altering the curvature of one or all of the surfaces.

A very small alteration will make a large difference in the optical properties, so that in general "figuring" is done merely by using the rouge polishing tool as an abrading tool, and causing it to alter the curves in the manner already suggested for grinding. There are other methods based on knocking squares out of the pitch-polisher so that some parts of the glass may be more abraded than others.

The "figuring" and polishing may be done by hand just like the grinding. There are machines, however, which can be made to execute the proper motions, and a polisher is set in such a machine, and the mechanical work done is by no means inconsiderable. In fact for surfaces above six inches in diameter few people are strong enough to work a polisher by hand owing to the intense adhesion between it and the exactly fitting glass surface.

Such is a general outline of the processes required to produce a lens or mirror. These processes will now be dealt with in much greater detail, and a certain amount of repetition of the above will unfortunately be necessary: the reader is asked to pardon this. It will also be advisable for the reader to begin by reading the whole account before he commences any particular operation. The reason for this is that it has been desirable to keep to the main account as far as possible without inserting special instructions for subsidiary operations, however important they may be; consequently it may not always be quite clear how the steps described are to be performed. It will be found, however, that all necessary information is really given, though perhaps not always exactly in the place the reader might at first expect.

54. All the discs that I have seen, come from the makers already roughly ground on the edges to a circular figure--but occasionally

the figure is very rough indeed — and in some cases, especially if small lenses have to be made, it is convenient to begin by cutting the glass discs out of glass sheet, which also may be purchased of suitable glass. To do this, the simplest way is to begin by cutting squares and then cutting off the corners with the diamond, the approximate circular figure being obtained by grinding the edges on an ordinary grindstone.

If the pieces are larger, time and material may be saved by using a diamond compass, i.e. an ordinary drawing compass armed with a diamond to cut circles on the glass, and breaking the superfluous glass away by means of a pair of spectacle-maker's shanks (Fig. 44), or what does equally well, a pair of pliers with soft iron jaws. With these instruments glass can be chipped gradually up to any line, whether diamond-cut or not, the jaws of the pincers being worked against the edge of the glass, so as to gradually crush it away.

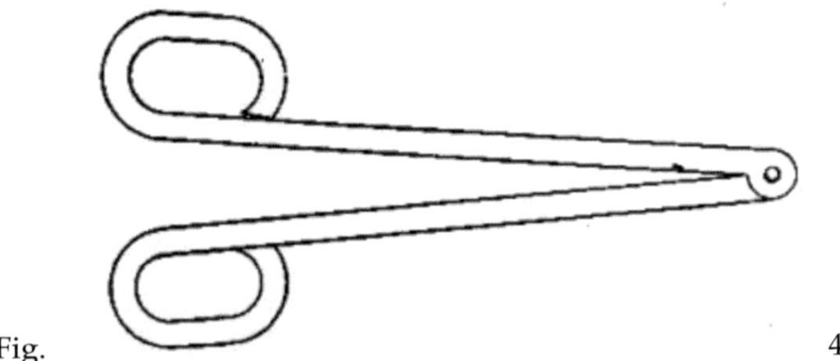

Fig. 44.

Assuming that the glass has been bought or made roughly circular, it must be finished on the lathe. For this purpose it is necessary to chuck it on an iron or hardwood chuck, as shown in Fig. 46. For a lens below say an inch in diameter, the centering cement may be used; but for a lens of a diameter greater than this, sufficient adhesion is easily obtained with Regnault's mastic, and its low melting point gives it a decided advantage over the shellac composition.

The glass may be heated gradually by placing it on the water bath, or actually in the water, and gradually bringing the water up to the boiling-point. The glass, being taken out, is rapidly wiped, and rubbed with a bit of waste moistened, not wet, with a little turpentine: its surface is then rubbed with a stick of mastic previously warmed so as to melt easily. The surface of the chuck being also warm, and covered with a layer of melted cement, it is applied to the glass. The lathe is turned slowly by hand, and the glass pushed gradually into the most central position; it is then pressed tight against the chuck by the back rest, a bit of wood being interposed for obvious reasons.

When all is cold the turning may be proceeded with. The quickest way is to use the method already described (i.e. actual turning by a file tool); but if the student prefers (time being no object), he may accomplish the reduction to a circular form very easily by grinding.

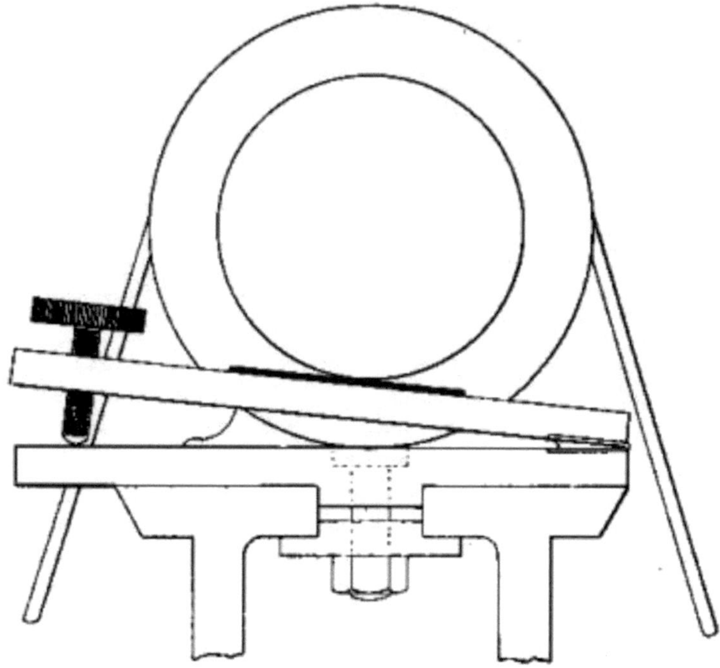

Fig. 45.

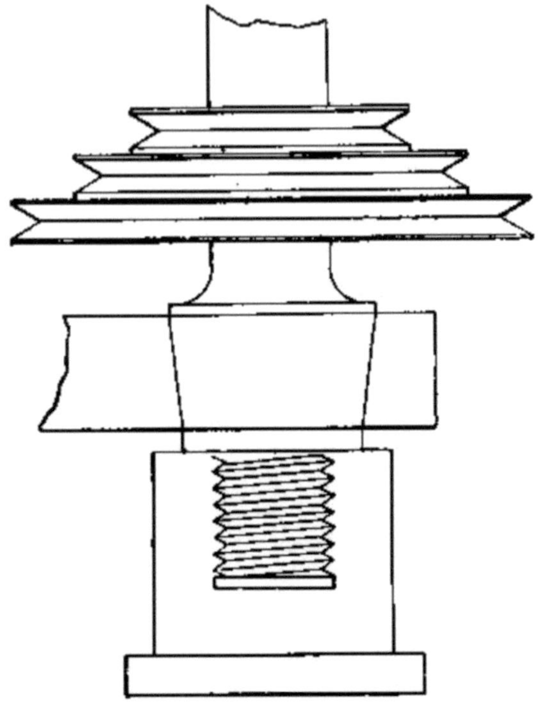

Fig. 46.

For this purpose he will require to make the following arrangements (Fig. 45). If the lathe has a slide rest, a piece of stout iron may be bent and cut so as to fit the tool rest, and project beneath the glass. The iron must be fairly rigid, for if it springs appreciably beneath the pressure of the glass, it will not grind the latter really round. The lathe may run rather faster than for turning cast iron of the same size. Coarse emery, passing through a sieve of 80 threads to the inch (run), may be fed in between the glass and iron, and the latter screwed up till the disc just grinds slightly as it goes round.

A beginner will generally (in this as in all cases of grinding processes) tend to feed too fast — no grinding process can be hurried. If a slide rest is not available, a hinged board, carrying a bit of iron, may (see Fig. 45) be arranged so as to turn about its hinge at the back of the lathe; and it may be screwed up readily enough by passing a long set-screw through the front edge, so that the point of the screw bears upon the lathe bed. I may add that emery behaves as if it were greasy, and it is difficult to wet it with clean water. This is easily got over by adding a little soap or alcohol to the water, or exercising a little patience.

A good supply of emery and water should be kept between the disc and the iron; a little putty may be arranged round the point of contact on the iron to form a temporary trough. In any case the resulting emery mud should on no account be thrown away, but should be carefully kept for further use. The process is complete when the glass is perfectly round and of the required diameter as tested by callipers.

55. The next step is to rough out the lens, and this may easily be done by rotating it more slowly, i.e. with a surface speed of ten feet per minute, and turning the glass with a hard file, as explained in 42. If it is desired to employ the slide rest, it is quicker and better to use a diamond tool — an instrument quite readily made, and of great service for turning emery wheels and the like, — a thing, in fact, which no workshop should be without. A bit of diamond bort, or even a clear though off-colour stone, may be employed.

An ordinary lathe tool is prepared by drawing down the tool steel to a long cone, resembling the ordinary practice in preparing a boring tool. The apex of the cone must be cut off till it is only slightly larger than the greatest transverse diameter of the diamond splinter. The latter may have almost any shape — a triangular point, one side of a three-sided prism is very convenient. A hole is drilled in the steel (which must have been well softened), only just large enough to allow the diamond to enter — if the splinter is thicker in the middle than at either end, so much the better — the diamond is fastened in position by squeezing the soft steel walls tightly down upon it. Personally I prefer to use a tool holder, and in this case generally mount the diamond in a bit of brass rod of the proper diameter; and instead of pinching in the sides of the cavity, I tin them, and set the diamond in position with a drop of soft solder.

Fig. 47.

In purchasing diamond bort, a good plan is to buy fragments that have been employed in diamond drilling, and have become too small to reset; in this case some idea as to the hardness of the bits may be obtained. Full details as to diamond tool-making are given in books on watch-making, and in Holtzapffell's great work on Mechanical Manipulation; but the above notes are all that are really necessary — it is, in fact, a very simple matter. The only advantage of using a diamond tool for glass turning is that one does not need to be always taking it out of the rest to sharpen it, which generally happens with hard steel, especially if the work is turned a little too fast.

I recommend, therefore, that the student should boldly go to work "free hand" with a hard file; but if he prefer the more formal method, or distrust his skill (which he should not do), then let him use a diamond point, even if he has the trouble of making it. When using a diamond it is not necessary to employ a lubricant, but there is some advantage in doing so.

The surface of the lens can be roughly shaped by turning to a template or pattern made by cutting a circular arc (of the same radius as the required surface) out of a bit of sheet zinc. Another very handy way of making templates of great accuracy is to use a beam compass (constructed from a light wooden bar) with a glazier's diamond instead of a pencil. A bit of thin sheet glass is cut across with this compass to the proper curvature--which can be done with considerable accuracy and the two halves of the plate, after breaking along the cut, are ground together with a view to avoiding slight local irregularities, by means of a little fine emery and water laid between the edges. In this process the glass is conveniently supported on a clean board or slate, and the bits are rubbed backwards and forwards against each other.

56. It is not very easy for a beginner to turn a bit of anything — iron, wood, or glass - with great accuracy to fit a template, and consequently time may be saved by the following procedure, applied as

soon as the figure of the template is roughly obtained. A disc of lead or iron, of the same diameter as the glass, and of approximately the proper curvature, is prepared by turning, and is armed with a handle projecting coaxially from the back of the disc. The glass revolving with moderate speed on the lathe, the lead tool, supplied with coarse emery and water, is held against it, care being taken to rotate the tool by the handle, and also to move it backwards and forwards across the disc, through a distance, say, up to half an inch; if it is allowed to overhang too much the edges of the glass disc will be overground. By the use of such a tool the glass can readily be brought up to the template.

The only thing that remains, so far as the description of this part of the process goes, is to give a note or two as to the best way of making the lead tools, and for this purpose the main narrative of processes must be interrupted. The easiest way is to make a set of discs to begin with. For this purpose take the mandrel out of the lathe, and place it nose downwards in the centre of an iron ring of proper diameter on a flat and level iron plate.

The discs are made by pouring lead round the screw-nose of the mandrel. This method, of course, leaves them with a hole in the centre; but this can be stopped up by placing the hot disc (from which the mandrel has been unscrewed) on a hot plate, and pouring in a sufficiency of very hot lead; or, better still, the mandrel can be supported vertically at any desired distance above the plate while the casting is being poured. Lead discs prepared in this way are easily turned so as to form very convenient chucks for brass work, and for use in the case now being treated, they are easily turned to a template, using woodturners' tools, which work better if oiled, and must be set to cut, not scrape.

If the operator does not mind the trouble of cutting a screw, or if he has a jaw chuck, the lead may be replaced by iron with some advantage.

The following is a neat way of making concave tools. It is an application of the principle of having the cutting tool as long as the radius of curvature, and allowing it to move about the centre of curvature. Place the disc of iron or lead on the lathe mandrel or in the chuck, and set the slide rest so that it is free to slide up or down the lathe bed. Take a bar of tool steel and cut it a little longer than the radius of curvature required. Forge and finish one end of the bar into a pointed turning tool of the ordinary kind. Measure the radius of curvature from the point of the tool along the bar, and bore a hole, whose centre is at this point, through the bar from the upper to the lower face. I regard the upper face as the one whose horizontal plane contains the cutting point when the tool is in use. Clamp a temporary back centre to the lathe bed, and let it carry a pin in the vertical plane through the lathe centres, and let this pin exactly fit the hole in the bar.

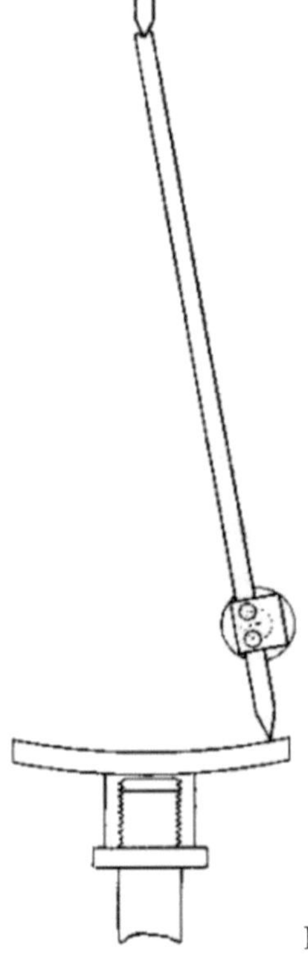

Fig. 48.

Place the "radius" tool in position for cutting, and let it be lightly held in the slide rest nearly at the cutting point, the centre of rotation of the pedestal (or its equivalent) passing through the central line of the bar. Then adjust the temporary back rest, so that the tool will take a cut. In the sketch the tool is shown swinging about the back centre instead of about a pin — there is little to choose between the methods unless economy of tool steel is an object. The tool must now be fed across the work. The pedestal must of course be free to rotate, and the slide rest to slip up and down the bed. In this way a better concave grinding tool can be made than would be made by a beginner by turning to a template — though an expert turner would probably carry out the latter operation so as to obtain an' accuracy of the same order, and would certainly do it in much less time than would be required in setting up the special arrangements here described.

On the other hand, if several surfaces have to be prepared, as in the making of an achromatic lens, the quickest way would be by the use of the radius tool, bored of course to work at the several radii required. I have tried both methods, and my choice would depend partly on the lathe at my disposal, and partly on the number of grinding tools that had to be prepared.

Having obtained a concave tool of any given radius, it is easily copied — negatively, so as to make a convex tool in the following manner. Adjust the concave tool already made on the back rest, so that if it rotated about the line of centres, it would rotate about its axis of figure.

Arrangements for this can easily be made, but of course they will depend on the detailed structure of the lathe. Use the slide rest as before, i.e. let it grasp an ordinary turning tool lightly, the pedestal being fixed, but the rest free to slide up or down the lathe bed. Push the back rest up till the butt of the turning tool (ground to a rounded point) rests against the concave grinding tool. If the diameter of the convex tool required be very small compared with the radius of curvature of the surface (the most usual case), it is only necessary to feed the cutting tool across to "copy" the concave surface sufficiently nearly.

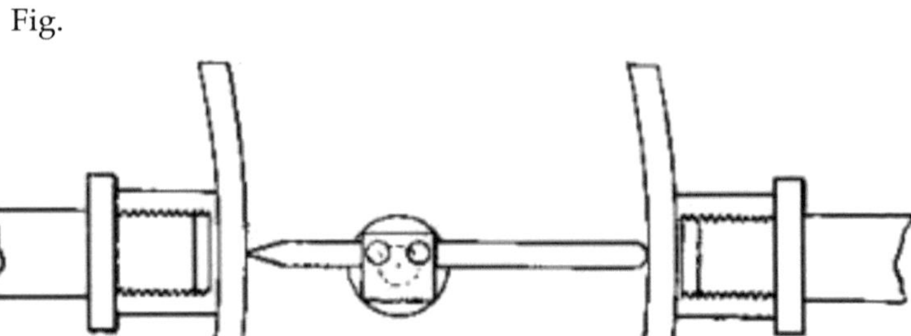

There seems no reason, however, why these methods should not be applied at once to the glass disc by means of a diamond point, and the rough grinding thus entirely avoided. I am informed that this has been done by Sir Henry Bessemer, but that the method was found to present no great advantage in practice. A reader with a taste for mechanical experimenting might try radius bar tools with small carborundum wheels rapidly driven instead of a diamond.

Enough has now been said to enable any one to prepare rough convex or concave grinding tools of iron or lead, and of the same diameter as the glass to be ground.

The general effect of the process of roughing the rotating lens surface is to alter the radius of curvature of both tool and glass; hence it is necessary to have for each grinding tool another to fit it, and enable it to be kept (by working the two together) at a constant figure. After a little practice it will be found possible to bring the glass exactly up to the required curvature as tested by template or spherometer. The art of the process consists in altering the shape of the grinding tool so as to take off the glass where required, as described in 53, and from this point of view lead has some advantages; (opinions vary as to the relative advantages of lead and iron tools for this purpose, however). The subsidiary grinding tool is not actually needed for this preliminary operation, but it has to be made some time with a view to further procedure, and occasionally is of service here.

57. 'The glass disc must be ground approximately to the proper curvature on each side before any fine grinding is commenced. It is precisely for this purpose that the previous turning of the disc is recommended, for it is easy to unmount and recentre a round object, but not so easy if the object have an indefinite shape. Using a cement which is plastic before it sets, the disc may be easily taken off the chuck and centred by a little handicraft, i.e. by rotating the lathe slowly and pushing the disc into such a position that it rotates about its axis. The grinding of the second surface is accomplished exactly as in the former case; of course on reversing the glass the chuck has to be slightly turned up to fit the convex or concave surface.

58. There is, however, one point of interest and importance — attention to which will save a good deal of useless labour afterwards. The glass must be ground in such a manner that the thickness at the edge is the same all round. In other words, the axes of figure of the two surfaces must coincide. This will be the case if the recentering has been accurately performed, and therefore no pains should be spared to see that it is exactly carried out. Any simple form of vernier gauge (such as Brown and Sharpe's vernier callipers) will serve to allow of a sufficiently accurate measurement of the edge thickness of the lens. If any difference of thickness is observed as the gauge moves round the edge, one or other of the surfaces must be reground. Of course the latitude of error which may be permitted depends so much on the final arrangements for a special finishing process called the "centering of the lens" — which will be described — that it is difficult to fix a limit, but perhaps one-thousandth of an inch may be mentioned as a suitable amount for a 2-inch disc. For rough work, of course; more margin may be admitted.

59. In a large shop I imagine that lenses of only two inches diameter would be ground in nests; or, in other words, a number would be worked at a time, and centering, even of a rough kind, would be left to the last; but this process will be treated hereafter. At present I shall assume that only one lens will be made at a time. Consequently we now enter on the stage of fine grinding by hand. A leaden pedestal, for the sake of stability, must be provided on which to mount the lens, so that the surface to be operated on may be nearly horizontal (Fig. 50). Before this can be done, however, fresh grind-

ing tools (two for each surface) must be properly prepared. After trying several plans I unhesitatingly recommend that all fine-grinding surfaces should be made of glass. This is easily done by taking two discs of lead, or iron, or slate, cut to a one-tenth inch smaller radius of curvature (in the case of a convex tool, and the opposite in the other case) than the lens surface (Fig. 51, A). On these, square bits of sheet glass, one-tenth of an inch thick, are to be cemented, so as to leave channels of about one-eighth of an inch between each bit of glass (Fig. 52, B). The "mastic" cement formerly described may be employed for this purpose.

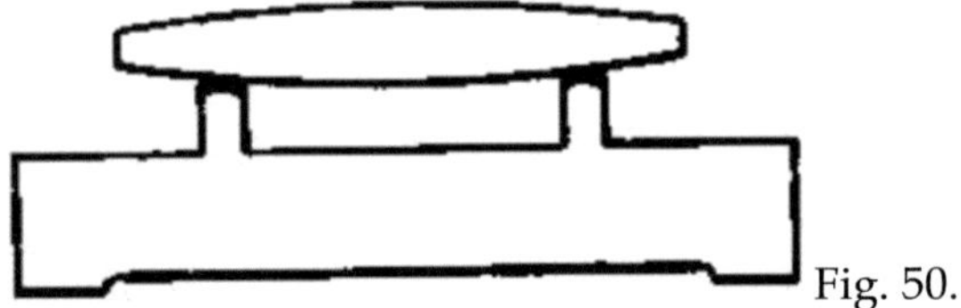
Fig. 50.

The bits of glass ought first to have their edges dressed smooth on the grind-stone. A convex and concave glass surface having been thus roughly prepared, they must be mounted in turn in the lathe, and brought to the proper curvature by grinding with the tools formerly employed and tested by the template or spherometer. It is well to control this process by means of a spherometer, so that the desired radius may be approximately reached. The two glass-grinding tools are then ground together by hand (see 53 and 61), the spherometer being employed from time to time to check the progress of the work. In general, if large circular sweeps are taken, greatly overhanging the side of the glass surface to be figured, both the upper and lower surfaces will be more ground at the edges, while in the opposite event the centre will be chiefly affected.

124

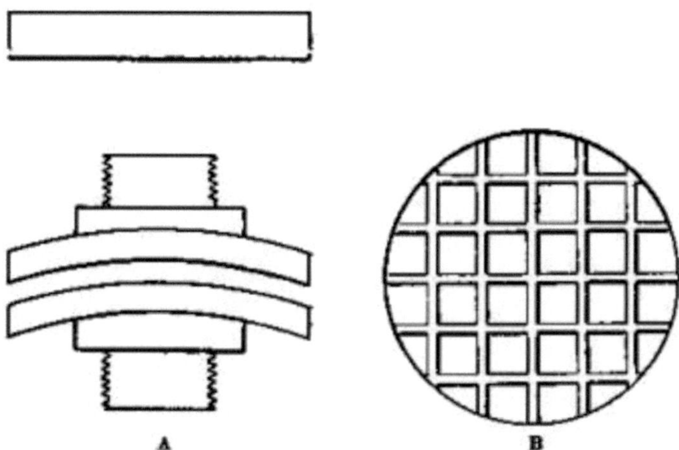

Fig. 51.

A spherometer capable of measuring a 2-inch surface may be procured, having a screw of, say, 50 threads to the inch, and a micrometer surface divided into 200 parts, each part easily capable of subdivision--into tenths or even twentieths. To get the full advantage of the spherometer it must screw exceedingly freely (i.e. must be well oiled with clock oil), and must not be fingered except at the milled head. If one of the legs is held by the fingers the expansion is sufficient to throw the instrument quite out of adjustment. The glass-grinding tools being brought to the proper figure, the next process is to transfer the same to the lens, and this is done by similar means, the fellow tool being used to correct the one employed in grinding the lens surface. Before the grade of emery is changed all three surfaces must agree, as nearly, at least, as the spherometer will show.

In order to prevent confusion the following summary of the steps already taken may be given. The discs of glass are first ground or turned so as to be truly circular. Four "tools" are made for each surface — a rough pair of iron or lead, and a finishing pair of iron, lead, or slate faced by glass squares. For a small lens the iron or lead backing may be used, for a large one the slate. The rough tools are used to give an approximate figure both to the lens and to the finishing tools.

The final adjustment is attained by grinding one of the glass-faced tools alternately upon the lens and upon the fellow glass-faced tool. The spherometer is accepted at all stages of the process as the final arbiter as to curvature. Some hints on the form of strokes used in grinding will be given later on (see 61). It suffices to state here that the object throughout is to secure uniformity by allowing both the work and the tool to rotate, and exercising no pressure by the fingers. The tool backing may weigh from one to two pounds for a 2-inch lens.

60. The tools and lens being all of the same curvature, the state of the surface is gradually improved by grinding with finer and finer emery. The best way of grading the emery is by washing it with clean water, and allowing the emery (at first stirred up with the water) to settle out. The longer the time required for this part of the process the finer will be the emery deposited. An ordinary bedroom jug is a very good utensil to employ during this process; a large glass jug is even better. The following grades will be found sufficient, though I daresay every operative's practice differs a little on this point.

1st grade. — Flour emery, with the grit washed out, i.e. allowed to stand for 2" (sec.) before being poured off.

2nd grade.--Stand 5" (secs.), settle in 1' (min.)

3rd grade. — Stand 1', settle in 10'.

4th grade. — Stand 10', settle in 60'.

It is generally advisable to repeat the washing process with each grade. Thus, selecting grade 2 for illustration, the liquor for grade 3 must be poured off without allowing any of the sediment to pass over with it. If any sediment at all passes, one has no security against its containing perhaps the largest particle in the jug. As soon as the liquor for No. 3 has been decanted, jug No. 2 is filled up again with clean water (filtered if necessary), and after standing 5" is decanted into jug No. 2b, the sediment is returned to jug No. 1, and the liquor, after standing 1', is transferred to jug No. 3.

The greatest care is necessary at each step of the operation to prevent "sediment" passing over with liquor. There is a little danger from the tendency which even comparatively large particles of emery have to float, in consequence of their refusing to get wet, and the emery worked up on the side of the jug is also a source of danger, therefore wipe the jug round inside before decanting.

In order to get a uniform grade stop the currents of water in the jug, which may work up coarse particles, by holding a thin bit of wood in the rotating liquid for a moment, and then gently withdrawing it in its own plane. These precautions are particularly necessary in the case of grades Nos. 2, 3, and 4, especially No. 4, for if a single coarse particle gets on the tool when the work has progressed up to this point it will probably necessitate a return to grinding by means of No. 2, and involve many hours' work.

The surface of the lens will require to be ground continuously with each grade till it has the uniform state of roughness corresponding to the grade in question. Two hours for each grade is about the usual time required in working such a lens as is here contemplated.

The coarser grades of emery may be obtained by washing ordinary flour of emery, but the finer ones have to be got from emery which has been used in the previous processes. It is not a good plan to wash the finer grades of emery out of the proceeds of very rough grinding say with anything coarser than flour of emery — as there is a danger of thereby contaminating the finer grades with comparatively coarse glass particles (owing to their lightness) and this may lead to scratching. If the finer grades are very light in colour, it may be inferred that a considerable portion of the dust is composed of glass, and this does no good. Consequently time may be saved by stirring up the light-coloured mass with a little hydrofluoric acid in a platinum capsule; this dissolves the finely divided glass almost instantaneously. The emery and excess of hydrofluoric acid may then be thrown into a large beaker of clean water and washed several times. Fine emery thus treated has much the same dark chocolate colour as the coarser varieties.

The operator should not wear a coat, and should have his arms bare while working with fine emery, for a workshop coat is sure to have gathered a good deal of dust, and increases the chances of coarse particles getting between the surfaces.

61. Details of the Process of Fine Grinding. —

A lens of the size selected for description is mounted as before mentioned on a leaden pedestal, and the operator places the latter on a table of convenient height in a room as free from dust as possible. Everything should be as clean as a pin, and no splashes of emery mud should be allowed to lie about. I have found it convenient to spread clean newspapers on the table and floor, and to wear clean linen clothes, which do not pick up dust. I have an idea that in large work-shops some simpler means of avoiding scratches must have been discovered, but I can only give the results of my own experience. I never successfully avoided scratches till I adopted the precautions mentioned.

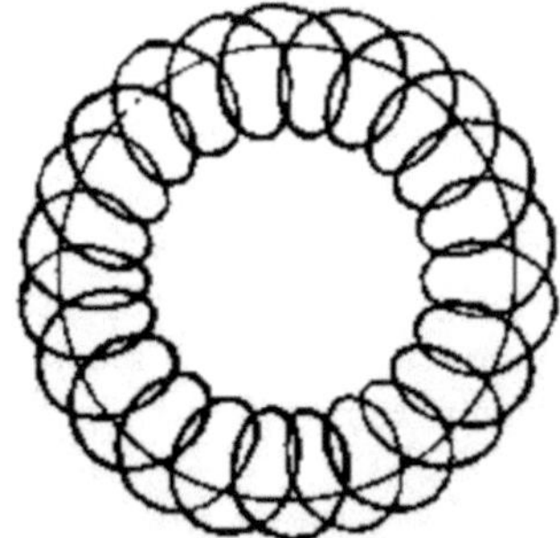

Fig. 52.

The left hand should be employed in rotating the pedestal either continuously (though slowly) or at intervals of, say, one minute. This point is rather important. Some operators require two hands to work the grinding tool, and in any case this is the safer practice. Under these circumstances the pedestal may be rotated through one-eighth or tenth of a revolution every three minutes, or thereabouts. The general motion given to the grinding tool should be a series of circular sweeps of about one-fourth the diameter of the glass disc, and gradually carried round an imaginary circle drawn on the surface of the lens and concentric with it (Fig. 52).

The tool may overhang the lens by a quarter of the diameter of the latter as a maximum. The circuit may be completed in from twelve to thirty sweeps. The grinding tool should be lightly held by the fingers and the necessary force applied parallel to the surface. The tool itself must be slowly rotated about its axis of figure. If the tool be lightly held, it will be found that it tends to rotate by itself. I say "tends to rotate," for if the tool be touching evenly all over the surface it will rotate in a direction opposite to the direction of the circular sweep. For instance, if the tool be carried round its looped path clockwise, it will tend to rotate about its own axis of figure counter-clockwise. If it touch more in the middle, this rotation will be increased, while if it touches more along the edge, the rotation will be diminished, or even reversed in an extreme case.

Every fifty sweeps or so the tool should be simply ground backwards and forwards along a diameter of the lens surface. This grinding should consist of three or four journeys to and fro along, say, eight different diameters. About one-quarter of the whole grinding should be accomplished by short straight strokes, during which the tool should only overhang about one-quarter of an inch. The object of the straight strokes is to counteract the tendency to a gradual accumulation of the emery in the centre, which results from the circular grinding.

A great deal of the art of the process consists in knowing how to work the tool to produce any given effect. For instance, if the lens requires to be ground down near the centre, the epicycloidal strokes must be nearly central; the tool must never overhang very much. If, on the other hand, it is the edges which require attention, these must be dealt with by wider overhanging strokes. The tool must be frequently tested on its fellow, and, indeed, ground upon it if any marked unevenness of action (such as that just described) is required for the lens. A check by spherometer will be applied at intervals according to the judgment of the operator, but, in any case, the fellow tool and lens should be kept at very nearly the same figure.

The emery should never be allowed to become anything like dry between the tool and the lens, for in some way (probably by capillary action increasing the pressure of the tool) this seems to lead to scratching and "rolling" of the emery. The channels in the glass tool between the squares are of the greatest importance in enabling the emery to distribute itself. Perhaps the best guide in enabling one to judge as to when it is time to wash off the emery and apply fresh is the "feel" of the tool; also when the mud gets light in colour we know that it is full of glass dust, and proportionately inoperative.

New emery may be put on, say, every five minutes, but no absolute rule can be given, for much depends on the pressure of the tool upon the lens. In the case considered a brass or lead, or even slate tool, of an inch, or even less, in thickness, will press quite heavily enough. In washing the lens and tool before new emery is introduced, a large enamelled iron bucket is very handy; the whole of the tool should be immersed and scrubbed with a nail-brush. The lens surface may be wiped with a bit of clean sponge, free from grit, or even a clean damp cloth.

When the time comes to alter the grade of emery, a fresh lot of newspapers should be put down, and tools, lens, and pedestal well washed and brushed by the nail-brush. The surfaces should be wiped dry by a fresh piece of rag, and examined for scratches and also for uniformity of appearance; a good opinion can be formed as to the fit of the surfaces by noting whether — and if so, to what degree — they differ in appearance from point to point when held so that the light falls on them obliquely.

It is necessary to exercise the greatest care in the washing between the application of successive grades of emery, and this will be facilitated if the edges of the glass squares were dressed on a grindstone before they were mounted. An additional precaution which may be of immense advantage is to allow the tool to dry between the application of successive grades of emery (of course, after it has been scrubbed), and then to brush it vigorously with a hat-brush. It sometimes happens that particles of mud which have resisted the wet scrubbing with the nail-brush may be removed by this method.

As my friend Mr. Cook informs me that his present practice differs slightly from the above, I will depart from the rule I laid down, and add a note on an alternative method.

Consider a single lens surface. This is roughed out as before by an iron tool, a rough fellow tool being made at the same time. The squares of glass are cemented to the roughing tool, and this is ground to the spherometer by means of the counterpart tool. The glass-coated tool is then applied to the lens surface and grinding with the first grade of emery commenced. The curvature is checked by the spherometer. Two auxiliary tools of, say, half the diameter of the lens, are prepared from slate, or glass backed with iron, and applied to grind down either the central part of the lens surface or tool surface, according to the indications of the spherometer. Any changes that may occur during grinding are corrected by these tools. The spherometer is accepted as the sole guide in obtaining the proper curvature. A slate backing is preferred for tools of any diameter over, say, 2 inches.

62. *Polishing.* —

After the surface has been ground with the last grade of emery, and commences to become translucent even when dry, the grinding may be considered to be accomplished, and the next step is the polishing. There are many ways of carrying out this process, and the relative suitability of these methods depends on a good many, so to speak, accidental circumstances. For instance, if the intention is to finish the polishing at a sitting, the polishing tool may be faced with squares of archangel — not mineral or coal-tar — pitch and brought to shape simply by pressing while warm against the face of the lens. A tool thus made is very convenient, accurate, and good, but it is difficult to keep it in shape for any length of time; if left on the lens it is apt to stick, and if it overhangs ever so little will, of course, droop at the edges.

On the whole, the following will be found a good and sufficient plan. The glass-grinding tool is converted into a polishing tool by pasting a bit of thin paper over its surface; a bit of woven letter paper of medium thickness with a smooth but not glazed surface does very well. We have found that what is called Smith's "21 lbs. Vellum Wove" is excellent. This is steeped in water till quite pliable and almost free from size. The glass tool is brushed over with a little thin arrowroot or starch paste, and the paper is laid upon it and squeezed down on the glass squares as well as possible; if the paper is wet enough and of the proper quality it will expand sufficiently to envelop the tool without creases, unless the curvature is quite out of the common.

This being accomplished, and the excess of water and paste removed, the face of the paper is (for security) washed with a little clean water and a bit of sponge, and, finally, the tool is slightly pressed on the lens so as to get the paper to take up the proper figure as nearly as possible. After the polishing tool has been thus brought to the proper figure, it is lifted off and allowed to dry slowly. When the paper is dry it may be trimmed round the edges so as not to project sensibly beyond the glass squares. The next step is to brush the surface over very carefully with polishing rouge (prepared as is described at the end of this section) by means of a hatbrush. When the surface of the paper is filled with rouge all excess must be removed by vigorous brushing.

Fig.

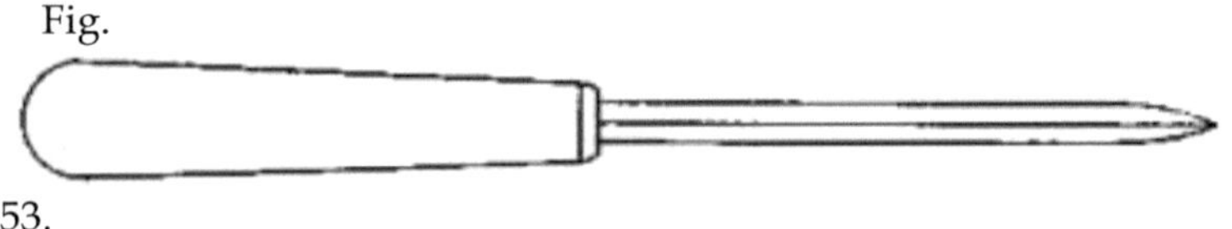

53.

The tool being placed on the lens, two or three strokes similar to those used in grinding may be taken, and the tool is then lifted off and examined. It will be found to be dotted with a few bright points, produced by the adhesion of glass at the places of contact. These points are then to be removed in the following manner. An old three-cornered file is ground on each side till the file marks disappear, and sharp edges are produced (Fig. 53). This tool is used as an ink eraser, and it will be found to scrape the paper of the polishing tool very cleanly and well.

The bright spots are the objects of attention, and they must be erased by the old file, and the polisher reapplied to the glass. A few strokes will develop other points, more numerous than before, and these in turn must be erased. The process is continued till the whole surface of the polishing tool is evenly covered with bright specks, and then the polishing may be proceeded with. The specks should not be more than about one-eighth of an inch apart, or the polishing will be irregular.

The operation of polishing is similar to that of grinding. A reasonable time for polishing a glass surface is twenty hours; if more time is required it is a sign that the fine grinding has not been carried far enough. The progress of the operation may be best watched by looking at the surface — not through it. For this purpose a good light is requisite. When the lens is dismounted it may be examined by a beam of sunlight in a dark room, under which circumstances the faintest signs of grayness are easily discernible.

It may be mentioned here that if the surface is in any way scratched the rouge will lodge in the scratches with great persistence, and an expert can generally tell from the appearance of scratches what kind of polishing powder has been employed.

The persistence with which rouge clings to a rough surface of glass is rather remarkable. Some glass polishers prefer to use putty powder as a polishing material, and it is sometimes said to act more quickly than rouge; from my rather limited experience I have not found this to be the case, but it may have merits that I do not know of. Is it possible that its recommendation lies in the fact that it does not render scratches so obtrusively obvious as rouge does?

Rouge is generally made in two or more grades. The softer grade is used for polishing silver, and is called jewellers' rouge. The harder grade, suitable for glass polishing, is best obtained from practical opticians (not mere sellers of optical instruments). I mean people like Messrs. Cook of York. Many years ago I prepared my own hard rouge by precipitating ferrous sulphate solution by aqueous ammonia, washing the precipitate, and heating it to a red heat. The product was ground up with water, and washed to get rid of large particles. This answered every purpose, and I could not find that it was in any way inferior to hard rouge as purchased. The same precipitate heated to a lower temperature is said to furnish a softer variety of rouge; at all events, it gives one more suitable for polishing speculum metal. Lord Rosso used rouge heated to a dull redness for this purpose.

Rouge, whether made or bought, should always be washed to get rid of grit. I ought to add that not the least remarkable fact about the polishing is the extraordinarily small quantity of the polishing material requisite, which suggests that the process of polishing is not by any means the same as that of exceptionally fine grinding. Is it possible that the chief proximate cause of the utility of rouge is to be sought in its curious property of adhering to a rough glass surface, causing it, so to speak, to drag the glass off in minute quantities, and redeposit it after a certain thickness has been attained on another part of the surface?

63. *Centering.* —

When a lens is ground and polished it will almost always happen that the axis of revolution of its cylindrical edge is inclined to the axis of revolution of its curved surfaces. Since in practice lenses have to be adjusted by their edges, it is generally necessary to adjust the edge to a cylinder about the axis of figure of the active surfaces. This is best done on a lathe with a hollow mandrel.. The lens is chucked on a chuck with a central aperture — generally by means of pitch or Regnault's mastic, or "centering" cement for small lenses — and a cross wire is fixed in the axis of revolution of the lathe, and is illuminated by a lamp. This cross wire is observed by an eye-piece (with cross wires only in the case of a convex lens, or a telescope similarly furnished in the case of a concave lens), also placed in the axis of rotation of the lathe.

Both cross wires are thus in the axis of revolution of the mandrel, and the distant one (B in the figure) is viewed through the lens and referred to the fixed cross wires at A. In general, as the lathe is rotated by turning the mandrel the image of the illuminated cross wires will be observed to rotate also. The lens is adjusted until the image remains steady on rotating the mandrel and it is to give time for this operation that a slow-setting cement is recommended. When the image remains stationary we know that the optical centre of the lens is in the axis of revolution, and that this axis is normal to both lens surfaces, i.e. is the principal axis of the lens, or axis of figure.

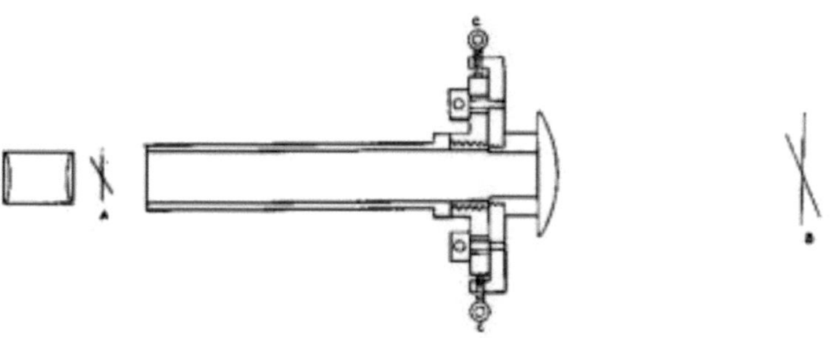

A much readier method, and one, in general, good enough for most purposes, is to put a candle on the end of the lathe-bed where the back centre generally is, and observe the images of the flame by reflection from both the lens surfaces. This method is very handy with small lenses; the mandrel is turned, and the lens adjusted by hand till the images are immovable. In both cases, of course, the edge of the lens is turned or ground till it is truly circular, the position of the lens remaining undisturbed on the chuck. If the edge gauge has been properly used in the earlier stages of figuring, it will be found that very little turning or grinding is requisite to produce a true centering.

The particular defect due to want of centering in a lens may be observed by using it as the objective of a telescope, and observing a star slightly out of focus. The interference fringes will not be concentric circles unless the lens is properly centred. I ought to say that I have not looked into the theory of this, but have merely taken it as a generally admitted fact. The diseases of lenses and the modes of treating them are dealt with in a book by Messrs. Cook of York, entitled *On the Adjustment and Testing of Telescopic Objectives*.

The final process of figuring will be dealt with later on (66 and 67), as it applies not only to lenses but to mirrors, prisms, etc. If the instructions given have been carefully carried out on a 2-inch lens, it should perform fairly well, and possibly perfectly, without any further adjustment of the glass.

64. *Preparation of Small Lenses, where great Accuracy is not of the first Importance. —*

Such lenses may generally be made out of bits of good plate or sheet glass, and are of constant use in the physical laboratory. They may be purchased so cheaply, however, that only those who have the misfortune to work in out-of-the-way places need be driven to make them.

Suitable glass having been obtained and the curves calculated from the index of refraction, as obtained by any of the ordinary methods applicable to plates (the microscope method, in general, is quite good enough), squares circumscribing the desired circles are cut out by the help of a diamond. *[Footnote:* Glazebrook and Shaw's Practical Physics, p. 383 (4th ed.).*]* The squares are roughly snipped by means of a pair of pliers or spectacle-maker's shanks. The rough circles are then mounted on the end of a brass or iron rod of rather greater diameter than the finished lenses are to possess. This mounting is best done by centering cement.

The discs are then dressed circular on a grindstone, the rod serving both as a gauge and handle. A sufficient number of these discs having been prepared, a pair of brass tools of the form shown in the sketch (Fig. 55), and of about the proper radius of curvature, are made. One of these tools is used as a support for the glass discs.

Fig.

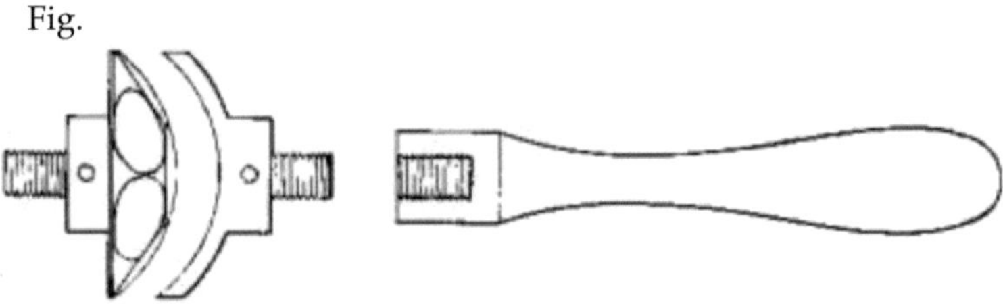

55.

A compass being set to scribe circles of the same diameter as the glass discs, centre marks are made on the surface of the appropriate tool, circles are drawn on this, and facets are filed or milled (for which the spiral head of the milling machine is excellent). In the case of concave supporting surfaces, i.e. in making concave lenses, I apprehend filing would be difficult, and the facets would have to be made by a rose cutter or mill; but if the discs are fairly round, then, in fact, no facets are required.

The facets being ready, the glass discs are cemented to them by centering cement, which may be used quite generally for small lenses. When the cutting of facets has been omitted on a concave surface, the best cement is hard pitch. The grinding tool is generally rather larger than the nest of lenses. Coarse and fine grinding is accomplished wholly on the lathe — the tool being rotated at a fair speed (see infra), and the nest of lenses moved about by its handle so as to grind all parts equally. It must, of course, be held anywhere except "dead on," for then the part round the axis would not get ground; this inoperative portion of the rotating tool must therefore be allowed to distribute its incapable efforts evenly over the nest of lenses.

Polishing is accomplished by means of the grinding tool, coated with paper and rouge as before; or the tool may be coated with very thin cloth and used with rouge as before — in this case the polishing goes on fastest when the surface of the cloth is distinctly damp. In working by this method, each grade of emery need only be applied from five to ten minutes. The glass does not appear to get scratched when the emery is changed, provided everything is well washed. A good polish may be got in an hour. The lathe is run as for turning brass of the same diameter as the tool.

One side of the lenses being thus prepared, they are reversed, and the process gone through for the other side in a precisely similar manner. *[Footnote: Unless the radius of curvature is very short and the lenses also convex, there is no necessity to recess the facets, provided hard pitch is used as the cement. See note on hard pitch.]* To save trouble, it is usual, to make such lenses of equal curvature on both faces; but of course this is a matter of taste.

Fig.

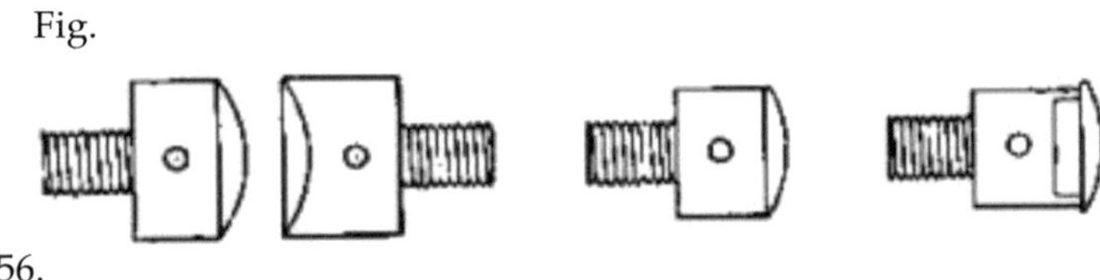

56.

For very common work, bits of good plate glass are employed, and the manufacturer's surface treated as flat (Fig. 56). In this way plano-convex lenses are easily and cheaply made. Finally the lenses have to be centred, an essential operation in this case. This is easily done by the reflection method — the edge being turned off by the file and kerosene and the centering cement being used in making the preliminary adjustment on the chuck. I presume a lens made in this way is worth about a shilling, so that laboratory manufacture is not very remunerative. Fig. 56 shows the method of mounting small lenses for lathe grinding, when only one lens is required. The tool is generally rotated in the lathe and the lens held against it.

65. Preparing Small Mirrors for Galvanometers. —

To get good mirrors for galvanometers, I have found the best plan is to grind and polish a large number together, on a disc perhaps 8 or 10 inches in diameter. I was led to this after inspecting and rejecting four ounces of microscope cover slips, a most wearisome process. That regular cover slips should be few and far between is not unlikely, seeing that they are made (by one eminent firm at least) simply by "pot" blowing a huge thin bulb, and then smashing it on the floor and selecting the fragments. As in the case of large mirrors, it is of course only necessary to grind one side of the glass, theoretically at all events. The objections to this course are:

1. A silver surface cannot, in my experience, be polished externally (on a minute object like a cover slip) to be anything like so bright as the silver surface next the glass; and,
2. if one side only is ground, it will be found that the little mirror hopelessly loses its figure directly it is detached from the support on which it has been worked. Conse-

quently, I recommend that these small mirrors should be ground and polished on both sides — enough may be made at one operation to last for a very long time.

A slate back is prepared of the same radius of curvature as it is desired to impart to the mirrors. Bits of thin sheet glass are then ground circular as described in the last section and cemented to this surface by the smallest quantity of clean archangel pitch, allowed to cool slowly and even to rest for a day before the work is proceeded with. The whole surface is then ground and polished as before.

The mirrors are now reversed, when they ought to nearly fit the tool (assuming that flats are being made, and the fellow tool in all other cases), and are recemented by pitch to the appropriate backing ground, and polished. If very excellent results are required, these processes may be preceded by a preliminary rough grinding of one surface, so that the little discs will "sit" exactly on the tool surface, and not run the risk of being strained by capillary forces in the pitch. We have always found this necessary for really good results.

On removing such mirrors from the backing, they generally, more or less lose their figure, becoming (in general fairly uniformly) more concave or convex. About 5 per cent of the mirrors thus prepared will be found almost perfect if the work has been well done, and the rest will probably be very fair, unless the diameter is very large as compared with the thickness. The best way of grinding and polishing such large surfaces (nests 10 inches in diameter) is on a grinding machine, such as will be described below. The polishing is best done by means of paper, as before described.

Having occasion to require hitherto unapproached lightness and optical accuracy in such mirrors, I got my assistant to try making them of fused quartz, slices being cut by a diamond wheel from a rod of that material. Chips of natural quartz were also obtained from broken "pebble" spectacles, and these were worked at the same time. The resulting mirrors were certainly superior to the best we could make from glass, but the labour of grinding was greater, and the labour of polishing less, than in the latter case. The pebble fragments gave practically as good mirrors as the fused slices. For the future it will be better always to make galvanometer mirrors from quartz crystals. These may be easily sliced, as will be described in 74. The slices are dressed on a grindstone according to instructions already given for small lenses.

The silvering of these mirrors is a point of great importance. After trying nearly every formula published, we have settled down to the following.

A solution of pure crystallised nitrate of silver in distilled water is made up to a strength of 125 grams of the salt per litre. This forms the stock solution and is kept in a dark bottle.

Let the volume of silvering liquor required in any operation be denoted by $4v$. The liquor is prepared as follows:

I. Measure out a volume v of the stock solution of silver nitrate, and calculate the weight of salt which it contains; let this be w. In another vessel dissolve pure Rochelle salt to the amount of $2.6w$, and make up the solution to the volume v. These two solutions are to be mixed together at a temperature of 55 C., the vessels with their contents being heated to this temperature on the water bath. After mixing the liquids the temperature is to be kept approximately constant for five minutes, after which the liquor may be cooled. The white precipitate which first forms will become gray or black and very dense as the liquid cools. If it does not, the liquor must be reheated to 55 C., and kept at that temperature for a few minutes and then again allowed to cool. The solution is in good order when all the precipitate is dense and gray or black and the liquor clear. The blacker and denser the precipitate the better is the solution. The liquor is decanted and filtered from the precipitate and brought up to the volume $2v$ by addition of some of the wash water.

II. Measure out a volume $0.118v$ of the stock solution into a separate vessel, and add to it a 5 per cent solution of ammonium hydrate, with proper precautions, so that the precipitate at first formed is all but redissolved after vigorous shaking. It is very important that this condition should be exactly attained. Therefore add the latter part of the ammonia very carefully. Make up the volume to $2v$.

Mix the solutions I. and II. in a separate vessel and pour the mixture into the depositing vessel. The surface to be silvered should face downwards, and lie just beneath the free surface of the liquid. Bubbles must of course be removed.

The silver deposit obtained in this manner is exceedingly white and, bright on the surface next to the glass, but the back is mat and requires polishing.

The detail of the process described above was worked out in my laboratory by Mr. A. Pollock, to whom my thanks are due.

This process gives good deposits when the solutions are freshly prepared, but the ammonia solution will not keep; The surfaces to be silvered require to be absolutely clean. The process is assisted by a summer temperature, say 70 Fahr., and possibly by the action of light. Six or seven hours at least are required for a good deposit; a good plan is to leave the mirrors in the bath all night. On removal from the bath the mirrors require to be well washed, and allowed to dry thoroughly in sun heat for several hours before they are touched.

Care should be taken not to pull the mirrors out of shape when they are mounted for the bath. A single drop of varnish or paint (a mere speck) on the centre will suffice to hold them. The back of the deposit requires to be varnished or painted as a rule to preserve the silver. All paints and varnishes thus applied tend to spoil the figure by expanding or contracting. On the whole, I think boiled linseed oil and white or red lead — white or red paint in fact — is less deleterious than other things I have tried. Shellac varnish is the worst.

Of course, the best mirror can be easily spoiled by bad mounting. I have tried a great number of methods and can recommend as fairly successful the following: — A little pure white lead, i.e. bought as pure as a chemical — not as a paint — is mixed with an equal quantity of red lead and made into a paste with a little linseed oil. I say a paste, not putty. A trace of this is then worked on to the back of the mirror at the centre as nearly as may be, and to this is attached the support. The only objection to this is that nearly a week is required for the paste to set. If people must use shellac let it be remembered that it will go on changing its shape for months after it has cooled (whether it has been dissolved in alcohol or not).

66. Preparation of Large Mirrors or Lenses for Telescopes. —

So much has been written on this subject by astronomers, generally in the *English Mechanic* and in the *Philosophical Transactions* for 1840, that it might be thought nothing could be added. I will only say here that the processes already described apply perfectly to this case; but of course I only refer to silver on glass mirrors. For any size over 6 inches in diameter, the process of grinding and polishing by hand, particularly the latter, will probably be found to involve too much labour, and a machine will be required. A description of a modification of Mr. Nasmyth's machine — as made by my assistant, Mr. Cook — will be found below.

There is no difficulty in constructing or working such a machine, and considered as an all round appliance, it possesses solid advantages over the simple double pulley and crank arrangement, which, however, from its simplicity deserves a note. Two pulleys, A and B, of about 18 inches diameter by 4 inches on the face, are arranged to rotate about vertical axes, and belted together. The shaft of one of these pulleys is driven by a belt in any convenient manner. Each pulley is provided on its upper surface with a crank of adjustable length carrying a vertical crank-pin.

Each crank-pin passes through a 3"X 2" wooden rod, say 3' 6" long, and these rods are pinned together at their farther extremities, and this pin carries the grinding or polishing tool, or rather engages loosely with the back of this tool which lies below the rod. It is clear that if the pulleys are of commensurable diameters, and are rigidly connected — say by belting which neither stretches nor slips — the polishing tool will describe a closed curve. If, however, the belt is arranged to slip slightly, or if the pulleys are of incommensurable diameters, the curve traced out by the grinding tool will be very complex, and in the case of the ratio of the diameters being incommensurable, will always remain open; for polishing purposes the consummation to be wished.

Mirror surfaces are ground spherical, the reduction to parabolic form being attained in the process of polishing. A very interesting account of the practice of dealing with very large lenses will be found in Nature, May 1886, or the Journal of the Society of Arts, same date (I presume), by Sir Howard Grubb. The author considers that the final adjustment of surfaces by "figuring" — of which more anon — is an art which cannot be learned by inspection, any more than a man could learn to paint by watching an artist. This is, no doubt, the case to some extent; still, a person wishing to learn how to figure a lens could not do better than take Sir Howard at his word, and spend a month at his works. Meanwhile the following remarks must suffice; it is not likely that anybody to whom these notes will be of service would embark on such large work as is contemplated by Sir Howard Grubb.

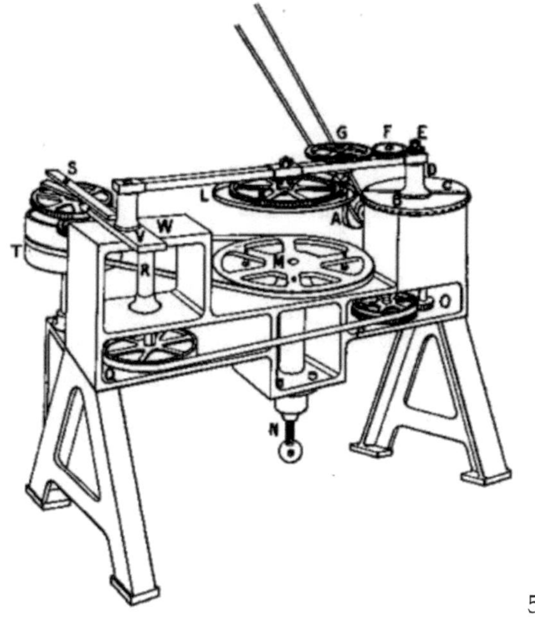

Fig. 57.

Description of Polishing Machine. Power is applied through belting to the speed cone A. By means of a bevel pinion rotation is communicated to the wheel D, which is of solid metal and carries a T-slot, C. A pedestal forming a crank-pin can be clamped so as to have any desired radius of motion by the screw E. A train of wheels E F G H K (ordinary cast lathe change wheels) communicate any desired ratio of motion to the tool-holder, which simply consists of

two pins projecting vertically downwards from the spokes of wheel
K.

These pins form a fork, and each prong engages in a corresponding hole in the back of the slate-grinding tool (not shown in figure). The connection with the tool is purposely loose. The wheel E, of course, cannot rotate about the crank-pin D. Provision for changing the ratio of tool rotation is achieved by mounting the wheels composing the train on pins capable of sliding along a long slot in the bar supporting them.

The farther end of this bar is caused to oscillate to and fro very slowly by means of an additional crank-pin S and crankshaft, the projecting face of the bed-plate W being placed so as to allow V to slide about easily and smoothly. Motion is communicated to this part of the system by means of gears at 0 and P, and a belt working from P to Q. Thus the vertical shaft R is set in motion and communicates by gears with S. A pulley placed on the axle of the wheel carrying the crank-pin S gives a slow rotation to the work which is mounted on the table M. A small but important feature is the tray L below the gear K. This prevents dirt falling from the teeth of the wheel on to the work. The motion of S is of course very much less than of B — say 100 times less. The work can be conveniently adjusted as to height by means of the screw N.

The machine must be on a steady foundation, and in a place as free from dust as possible. Though it looks complicated it is quite straight-forward to build and to operate.

It is explained in Lord Rayleigh's article on Optics in the *Encyclopaedia Britannica* that a very minute change in the form of the curvature of the surface of a lens will make a large difference in the spherical aberration. This is to be expected, seeing that spherical aberration is a phenomenon of a differential sort, i.e. a measure of the difference between the curvature actually attained, and the theoretical curvature at each point of the lens, for given positions of point and image. Sir H. Grubb gives an illustration of the minuteness of the abrasion required in passing from a curve of one sort to a curve of another, say from a spherical to a parabolic curve, consequently the process of figuring by the slow action of a polishing tool becomes quite intelligible. In making a large mirror or lens all the processes hitherto described under grinding and polishing, etc., have to be gone through and in the manner described, and when all this is accomplished the final process of correcting to test commences. This process is called figuring.

67. Of the actual operation of this process I have no personal knowledge, and the following brief notes are drawn from the article by Sir H. Grubb, from my assistant's (Mr. Cook) experience, and from a small work On the Adjustment and Testing of Telescopic Objectives, by T. Cook and Sons, Buckingham Works, York (printed by Ben Johnson and Co., Micklegate, York). This work has excellent photographs of the interference rings of star images corresponding to various defects. It must be understood that the following is a mere sketch. The art will probably hardly ever be required in laboratory practice, and those who wish to construct large telescopes should not be above looking up the references.

The process is naturally divided for treatment into two parts.

(1) The detection of errors, and the cause of these errors.

(2) The application of a remedy.

(1) A lens, being mounted with its final adjustments, is turned on to a star, which must not be too bright, and should be fairly overhead. The following appearances may be noted:-

A. In focus, the star appears as a small disc with one or two rings round it; inside and outside of the focus the rings increase in number, are round, concentric with the disc, and the bright and dark rings are apparently equally wide. The appearance inside the focus exactly resembles that outside when allowance is made for chromatic effects. Conclusion: objective good, and correctly mounted.

B. The rings round the star in focus are not circular, nor is the star at the centre of the system. In bad cases the fringes are seen at one side only. Effects exaggerated outside and inside the focus. Conclusion: the lens is astigmatic, or the objective is not adjusted to be co-axial with the eyepiece.

C. When in focus the central disc is surrounded by an intermittent diffraction pattern, i.e. for instance the system of rings may appear along, and near, three or more radii. If these shift when the points of support of the lens are shifted, flexure may be suspected.

D. On observing inside and outside the focus, the rings are not equally bright and dark. This may be due to uncorrected spherical aberration, particularly to a fault known as "zonal aberration," where different zones of the lens have different foci, but each zone has a definite focus.

E. Irregular diffraction fringes point to bad annealing of the glass. This may be checked by an examination of the lens in polarised light.

F. If the disc appear blurred and coloured, however the focus be adjusted, incomplete correction for chromatic aberration is inferred. If in addition the colouring is unsymmetrical (in an extreme case the star disc is drawn out to a coloured band), want of centering is to be inferred. This will also show itself by the interference fringes having the characteristics described in C.

(2) The following steps may be taken in applying a remedy:

A. The adjusting screws of the cell mounting the object glass may be worked until the best result is attained; this requires great care and patience. Any errors left over are to be attributed to other causes than the want of collinearity of the axes of object glass and eyepiece.

B. Astigmatism is detected by rotating the object glass or object glass cell. If the oval fringes still persist and the longer axis follows the lens, astigmatism may be inferred. Similarly, by rotating one lens on the other, astigmatism, or want of centering (quite a different thing) may be localised to the lens.

C. The presence of flexure may be confirmed by altering the position of the points of support with respect to the eyepiece, the lens maintaining its original position. The addition of more points of support will in general reduce the ill effects. How to get rid of them I do not know; they are only serious with large lenses.

D. Spherical aberration may be located by using stops and zonal screens, and observing the effect on the image. Sir H. Grubb determines whether any point on the lens requires to be raised or lowered, by touching the glass at that point with a warm hand or cooling it by ether. The effects so produced are the differential results of the change of figure and of refractive index. By observing the effect of the heating or cooling of any part, the operator will know whether to raise or lower that part, provided that by a suitable preliminary experiment he has determined the relation between the effect produced by the change of figure, and that due to the temperature variation of the refractive index. In general it is sufficient to consider the change of shape only and neglect the change in refractive power.

E. Marked astigmatism has never been noticed by me, but I should think that the whole lens surface would require to be repolished or perhaps reground in this case.

F. To decide in which surface faults exist is not easy. By placing a film of oil between the two surfaces nearly in contact these may be easily examined. Thus a mixture of nut and almond oil of the right proportion, to be found by trial, for the temperature, will have the same refractive index as the crown glass, and will consequently reduce any errors of figure in the interior crown surface, if properly applied between the surfaces. Similarly the interior of the flint surface may have its imperfections greatly reduced in effect by using almond oil alone, or mixed with bisulphide of carbon. The outer surfaces, I presume, must be examined by warming or cooling over suitable areas or zones.

The defects being detected, a matter requiring a great deal of skill and experience according to Sir H. Grubb, the next step is to remedy them; and the remedial measures as applied to the glass constitute the process of figuring. There are two ways of correcting local defects, one by means of small paper or pitch covered tools, which according to Sir H. Grubb is dangerous, and according to the experience of Mr. Cook, and I think of many French opticians, safe and advantageous.

Pitch polishing tools are generally used for figuring. They are made by covering a slate backing with squares of pitch. The backing is floated with pitch say one-eighth of an inch thick. This is then scored into squares by a hot iron rod. The tool, while slightly warm, is laid upon the lens surface, previously slightly smeared with dilute glycerine, until the pitch takes the figure of the glass. The polishing material is rouge and water. Small tools are applied locally, and probably can only be so applied with advantage for grave defects.

The other method is longer and probably safer. It consists in furnishing the polishing tool with squares of pitch as before. These being slightly warm, the lens is placed upon them so that they will flow to the exact figure also as before. I presume that the lens is to be slightly smeared with glycerine, or some equivalent, to keep the pitch from sticking. The squares are most thickly distributed where the abrasion is most required, i.e. less pitch is melted out by the iron rod. This may be supplemented by taking advantage of differences of hardness of pitch, making some squares out of harder, others out of softer pitch. The aim is to produce a polishing tool which will polish unequally so as to remove the glass chiefly from predetermined parts of the lens surface. The tool is worked over the surface of the lens by the polishing machine, and part of the art consists in adjusting the strokes to assist in the production of the local variations required.

A source of difficulty and danger lies in the fact that the pitch squares are rarely of the same hardness, so that some abrade the glass more rapidly than others. This is particularly likely to occur if the pitch has been overheated. *[Footnote: When pitch is heated till it evolves bubbles of gas its hardness increases with the duration of the process.]* The reader must be good enough to regard these remarks as of the barest possible kind, and not intended to convey more than a general idea of the nature of the process of figuring.

68. A few remarks on cleaning lenses will fittingly close this part of the subject. There is no need to go beyond the following instructions given by Mr. Brashear in Popular Astronomy, 1894, which are reproduced here verbatim.

"The writer does not advise the use of either fine chamois skin, tissue paper, or an old soft silk handkerchief, nor any other such material to wipe the lenses, as is usually advised. It is not, however, these wiping materials that do the mischief, but the dust particles on the lenses, many of them perhaps of a silicious nature, which are always harder than optical glass, and as these particles attach themselves to the wiping material they cut microscopic or greater scratches on the surfaces of the objective in the process of wiping.

"I write this article with the hope of helping to solve this apparently difficult problem, but which in reality is a very simple one.

"Let us commence by taking the object glass out of its cell. Take out the screws that hold the ring in place, and lift out the ring. Placing the fingers of both hands so as to grasp the objective on opposite sides, reverse the cell, and with the thumbs gently press the objective *squarely* out of the cell on to a book, block of wood, or anything a little less in diameter than the objective, which has had a cushion of muslin or any soft substance laid upon it. One person can thus handle any objective up to 12 inches in diameter.

"Before separating the lenses it should be carefully noted how they were put together with relation to the cell, and to one another, and if they art not marked they should be marked on the edges conspicuously with a hard lead pencil, so that when separated they may be put together in the same way, and placed in the same relation to the cell. With only ordinary precaution this should be an easy matter.

"Setting the objective on edge the two lenses may be readily separated.

"And now as to the cleaning of the lenses. I have, on rare occasions, found the inner surfaces of an object glass covered with a curious film, not caused directly by moisture but by the apparent oxidation of the tin-foil used to keep the lenses apart. "A year or more ago a 7-inch objective made by Mr. Clark was brought to me to clean. It had evidently been sadly neglected. The inside of the lenses were covered with such a film as I have mentioned, and I feared the glass was ruined. When taken apart it was found that the tin-foil had oxidised totally and had distributed itself all over the inner surfaces. I feared the result, but was delighted to find that nitric acid and a tuft of absorbent cotton cut all the deposit off, leaving no stains after having passed through a subsequent washing with soap and water.

"I mention this as others may have a similar case to deal with.

"For the ordinary cleaning of an objective let a suitable sized vessel, always a wooden one, be thoroughly cleaned with soap and water, then half filled with clean water about the same temperature as the glass. Slight differences of temperature are of no moment. Great differences are dangerous in large objectives.

"I usually put a teaspoonful of ammonia in half a pail of water, and it is well to let a piece of washed 'cheese cloth' lie in the pail, as then there is no danger if the lens slips away from the hand, and, by the way, every observatory, indeed every amateur owning a telescope, should have plenty of 'cheese cloth' handy. It is cheap (about 3 cts. per yard) and is superior for wiping purposes to any 'old soft silk handkerchief,' chamois skin, etc. Before using it have it thoroughly washed with soap and water, then rinsed in clean water, dried and laid away in a box or other place where it can be kept clean. When you use a piece to clean an objective throw it away, it is so cheap you can afford to do so.

"If the lenses are very dirty or 'dusty,' a tuft of cotton or a camel's-hair brush may be used to brush off the loose material before placing the lenses in the water, but no pressure other than the weight of the cotton or brush should be used. The writer prefers to use the palms of the hand with plenty of good soap on them to rub the surfaces, although the cheese cloth and the soap answers nicely, and there seems to be absolutely no danger of scratching when using the hands or the cheese cloth when plenty of water is used; indeed when I wish to wipe off the front surface of an objective in use, and the lens cannot well be taken out, I first dust off the gross particles and then use the cheese cloth with soap and water, and having gone over the surface gently with one piece of cloth, throw it away and take another, perhaps a third one, and then when the dirt is, as it were, all lifted up from the surface, a piece of dry cheese cloth will finish the work, leaving a clean brilliant surface, and no scratches of any kind.

"In washing large objectives in water I generally use a 'tub' and stand the lenses on their edge. When thoroughly washed they are taken out and laid on a bundle of cheese cloth and several pieces of the same used to dry them.

"I think it best not to leave them to drain dry; better take up all moisture with the cloth, and vigorous rubbing will do no harm if the surfaces have no abrading material on them. I have yet to injure a glass cleaned in this way.

"This process may seem a rather long and tedious one, but it is not so in practice, and it pays.

"In some places objectives must be frequently cleaned, not only because they become covered with an adherent dust, but because that dust produces so much diffused light in the field as to ruin some kinds of telescope work. Mr. Hale of the Kenwood Observatory tells me he cannot do any good prominence photography unless his objective has a clean surface; indeed every observer of faint objects or delicate planetary markings knows full well the value of a dark field free from diffused light. The object-glass maker uses his best efforts to produce the most perfect polish on his lenses, aside from the accuracy of the curves, both for high light value and freedom from diffused light in the field, and if the surfaces are allowed to become covered with dust, his good work counts for little.

"If only the front surface needs cleaning, the method of cleaning with cheese cloth, soap and water, as described above, answers very well, but always throw away the first and, if necessary, the second cloth, then wipe dry with a third or fourth cloth; but if the surfaces all need cleaning I know of no better method than that of taking the objective out of its cell, always using abundance of soap and water, and keep in a good humor."

69. The Preparation of Flat Surfaces of Rock Salt. —

The preliminary grinding is accomplished as in the case of glass, except that it goes on vastly faster. The polishing process is the only part of the operation which presents any difficulty. The following is an extract from a paper on the subject, by Mr. J. A. Brashear, Pittsburg, Pa., U.S.A., from the Proceedings of the American Association for the Advancement of Science, 1885. Practically the same method was shown me by Mr. Cook some years earlier, so that I can endorse all that Mr. Brashear says, with the following exceptions. We consider that for small salt surfaces the pitch is better scored into squares than provided with the holes recommended by Mr. Brashear.

Mr. Brashear's instructions are as follows. After alluding to the difficulty of drying the polished salt surface — which is of course wet — Mr. Brashear says:-

"Happily I have no trouble in this respect now, and as my method is easily carried out by any physicist who desires to work with rock salt surfaces, it gives me pleasure to explain it. For polishing a prism I make an ordinary pitch bed of about two and one-half or three times the area of the surface of the prism to be polished. While the pitch is still warm I press upon it any approximately flat surface, such as a piece of ordinary plate glass. The pitch bed is then cooled by a stream of water, and conical holes are then drilled in the pitch with an ordinary counter sink bit, say one-quarter of an inch in diameter, and at intervals of half an inch over the entire surface. This is done to relieve the atmospheric pressure in the final work. The upper surface of the pitch is now very slightly warmed and a true plane surface (usually a glass one, prepared by grinding and polishing three surfaces in the ordinary way, previously wetted) is pressed upon it until the pitch surface becomes an approximately true plane itself. Fortunately, moderately hard pitch retains its figure quite persistently through short periods and small changes of temperature, and it always pays to spend a little time in the preparation of the pitch bed.

"The polisher being now ready, a very small quantity of rouge and water is taken upon a fine sponge and equally distributed over its surface. The previously ground and fined salt surface (this work is done the same as in glass working) is now placed upon the polisher and motion instantly set up in diametral strokes. I usually walk around the polisher while working a surface. It is well to note that motion must be constant, for a moment's rest is fatal to good results, for the reason that the surface is quickly eaten away, and irregularly so, owing to the holes that are in the pitch bed. Now comes the most important part of this method. After a few minutes' work the moisture will begin to evaporate quite rapidly. No new application of water is to be made, but a careful watch must be kept upon the pitch bed, and as the last vestige of moisture disappears the prism is to be slipped off the polisher in a perfectly horizontal direction, and if the work has been well done, a clean, bright, and dry surface is the result. The surface is now tested by the well-known method of interference from a perfect glass test plate (see Fig. 178).

"If an error of concavity presents itself the process of polishing is gone over again, using short diametral strokes. If the error is one of convexity, the polishing strokes are to be made along the chords, extending over the edge of the polisher. The one essential feature of this method is the fact that the surface is wiped dry in the final strokes, thus getting rid of the one great difficulty of pitch polishing, a method undoubtedly far superior to that of polishing on broadcloth. If in the final strokes the surface is not quite cleaned I usually breathe upon the pitch bed, and thus by condensation place enough moisture upon it to give a few more strokes, finishing just the same as before. In ten minutes I have polished prisms of rock salt in this manner that have not only shown the D line double, but Professor Langley has informed me that his assistant, Mr. Keeler (J. E.), has seen the nickel line clearly between the D lines. This speaks for the superiority of the surfaces over those polished on broadcloth.

"In polishing prisms I prefer to work them on top of the polisher, as they can be easily held, but as it is difficult to hold lenses or planes in this way without injuring the surfaces, I usually support them in a block of soft wood, turned so as to touch only at their edges, and work the polisher over them. Though it takes considerable practice to succeed at first, the results are so good that it well repays the few hours' work it requires to master the few difficulties it presents."

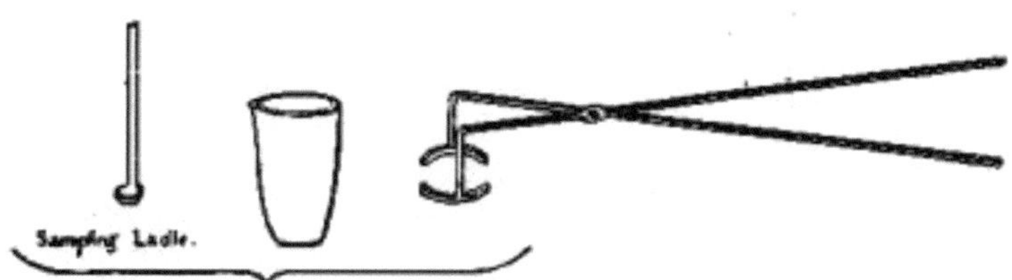

Fig. 58.

70. Casting Specula for Mirrors. —

According to Sir H. Grubb (loc. cit.) the best alloy is made of four atoms of copper and one of tin; this gives by weight, copper 252, tin 117.8.

The copper is melted first in a plumbago crucible; the tin is added gradually. Of course, in the process of melting, even though a little fine charcoal be sprinkled over the copper, some loss of that metal will occur from oxidation. It is convenient in practice, therefore to reserve a portion of the tin and test the contents of the crucible by lifting a little of the alloy out and examining it.

The following indications may be noted: When the copper is in excess the tint of the alloy is slightly red, and the structure, as shown at a fractured surface, is coarsely crystalline. As the proper proportions are more nearly attained, the crystalline structure becomes finer, the colour whiter, and the crystals brighter. The alloy is ready for use when the maximum brightness is attained and the grain is fine.

If too much tin be added, the lustre diminishes. The correct proportion is, therefore, attained when a further small addition of tin produces no apparent increase of brightness or fineness of grain. About three-quarters of the tin may be added at first, and the other quarter added with testing as described. The alloy is allowed to cool until on skimming the surface the metal appears bright and remains so without losing its lustre by oxidation for a sensible time; it will still be quite red-hot.

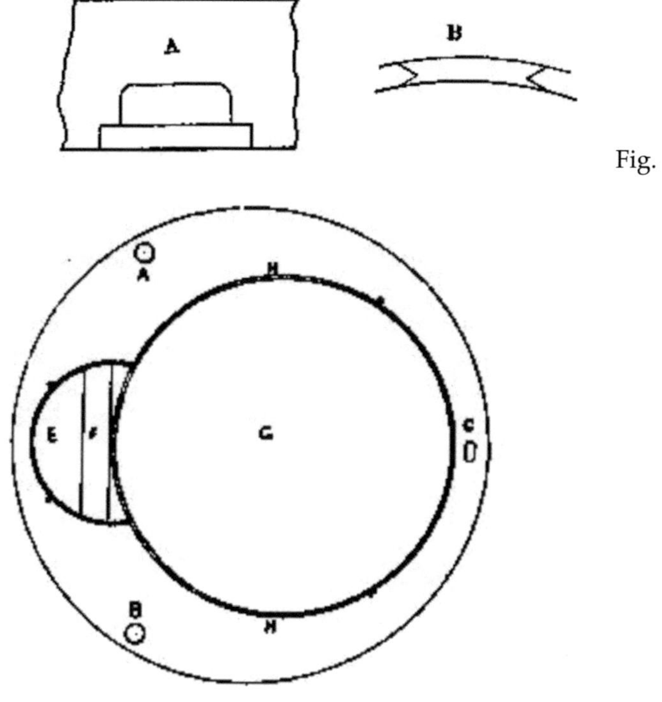

Fig. 59. Fig. 60.

As the speculum alloy is too difficult to work with ordinary tools, it is best to cast the speculum of exactly the required shape and size. This is done by means of a ring of iron turned inside (and out) and on one edge. This ring is laid on a plate of figured iron, and before the metal is poured the plate (G) (Figs 59 and 61) is heated to, say, 300 C. In order to avoid the presence of oxide as far as possible, the following arrangements for pouring are made. A portion of the lower surface of the ring is removed by radial filing until a notch equal to, say, one-twentieth of the whole circumference is produced. This is cut to an axial depth of, say, half an inch.

A bar of iron is then dovetailed loosely into the notch (Fig. 60, B), so that it will rest on the iron plate, and half fill the notch. The aperture thus left forms the port of ingress for the hot metal (see Fig. 61,

M). A bit of sheet iron is attached to the upper surface of the ring, and lies as a sort of flap, shaped like a deep shovel, against the outside of the ring overhanging the port (Figs. 59 and 61 at F). This flap does not quite reach the iron plate, and its sides are bent so as to be in contact with the ring. A portion of a smaller ring is then applied in such a manner as to form a pouring lip or pool on the outside of the main ring at E, and the metal can only get into the main ring by passing under the edge of the flap and up through the port. This forms an efficient skimming arrangement. The process of casting is carried out by pouring steadily into the lip.

To avoid air bubbles it is convenient to cause the metal to spread slowly over the chill, and Mr. Nasmyth's method of accomplishing this is shown in the figure (61). The chill rests on three pins, A B C (Figs. 59 and 61). Before pouring begins the chill is tilted up off C by means of the counterpoise D, which is insufficient to tilt it after the speculum is poured. It is important that the chill should be horizontal at the close of the operation, in order that the speculum may be of even thickness throughout. This is noted by means of levels placed on the ring (at K for instance).

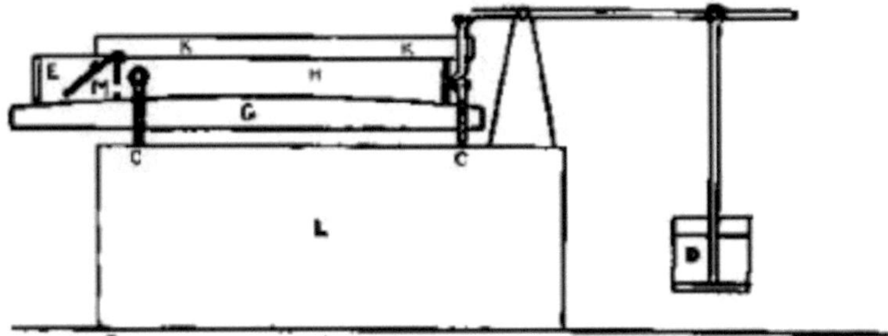

Fig. 61.

This apparatus may appear unnecessarily complex, but it is worth while to set it up, for it makes the operation of casting a speculum fairly certain. If the metal is at the right temperature it will form a uniformly liquid disc inside the ring. The mass sets almost directly, and as soon as this occurs it is pushed to the edge of the plate and the metal in the lip broken off by a smart upward tap with a hammer. The dovetailed bit of iron is knocked downwards and falls off, and the ring may then be lifted clear of the casting. The object of the dovetail will now be understood, for without it there is great risk of breaking into the speculum in knocking the "tail" off.

A box of quite dry sawdust is prepared in readiness for the process of annealing before the speculum is cast. The box must be a sound wooden or metal box, and must be approximately air-tight. For a speculum a foot in diameter the box must measure at least 3 feet both ways in plan, and be 2 feet 6 inches deep. Half the sawdust is in the box and is well pressed down so as to half fill it. The other half must be conveniently ready to hand. As soon as possible after casting, the speculum is thrown into the box, covered over with the sawdust, and the lid is put on.

The object in having the box nearly air-tight is to avoid air-currents, which would increase the rate of cooling. A speculum a foot in diameter may conveniently take about three days to anneal, and should be sensibly warm when the box is opened on the fourth day. For larger sizes longer times will be required. We will say that the sawdust thickness on each side must be proportional to the dimensions of the speculum, or may even increase faster with advantage if time is of no moment.

The process of annealing may be considered successful if the disc does not fly to pieces in working; it is to be worked on the chilled side. The object of giving the chill the approximate counterpart form will now appear; it saves some rough grinding, and causes the finished surface to be more homogeneous than it would be if the centre were sunk by grinding through the chilled surface.

In 1889 I learned from Mr. Schneider, Professor Row-land's assistant at Baltimore, that in casting specula for concave gratings a good deal of trouble had been saved by carrying out the operation in an atmosphere consisting mostly of coal gas. It was claimed that in this way the presence of specks of oxide was avoided. I did not see the process in operation, but the results attained are known and admired by all experimenters.

71. Grinding and polishing Specula. —

The rough grinding is accomplished by means of a lead tool and coarse emery; the size of grain may be such as will pass a sieve of 60 threads to the inch. The process of grinding is quite similar to that previously described, but it goes on comparatively quickly. The rough grinding is checked by the spherometer, and is interrupted when that instrument gives accordant and correct measurements all over the surface.

The fine grinding may be proceeded with by means of a glass-faced tool as before described, or the labour may be reduced in the following manner. A slate tool, which must be free from green spots (a source of uneven hardness), is prepared, and this is brought nearly to the curvature of the roughly ground speculum, by turning or otherwise. It is finished on the speculum itself with a little flour of emery. The fine grinding is then carried on by means of slate dust and water, the slate tool being the grinder. The tool is, of course, scored into squares on the surface.

If the casting process has been carried out successfully, the rough grinding may take, say six hours, and the fine grinding say thirty hours for a disc a foot in diameter. The greatest source of trouble is want of homogeneity in the casting, as evidenced by blowholes, etc. In general, the shortest way is to discard the disc and start afresh if there is any serious want of perfection in the continuity or homogeneity of the metal.

Fig.

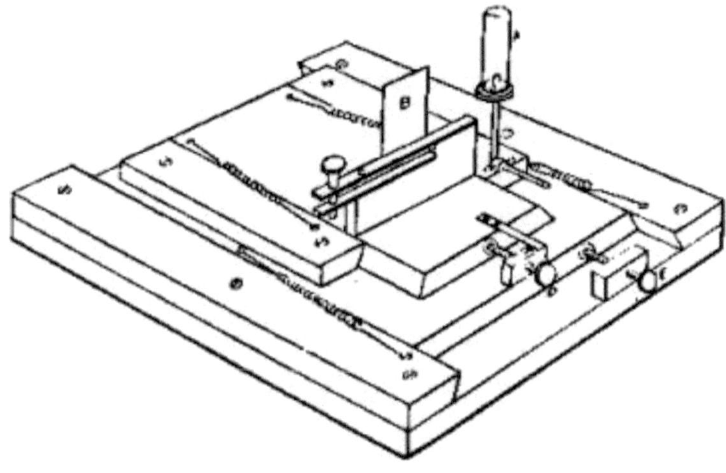

62.

The finely ground surface must, of course, be apparently correct in so far as a spherometer (with 3 inches between the legs for a disc 1 foot in diameter) will show. Polishing and figuring are carried out simultaneously. Half an hour's polishing with a slate-backed pitch tool and rouge and water will enable an optical test to be made. The most convenient test is that of Foucault, a simple appliance for the purpose being shown in the figure (62). It essentially consists of a small lamp surrounded by an opaque chimney (A) through which a minute aperture (pin-hole) is made. A small lens may be used, of very great curvature, or even a transparent marble to throw an image of the flame on the pin-hole.

A screen (B) is placed close to the source, and is provided with a rocking or tilting motion (C) in its own plane. The source and screen are partly independent, and each is provided with a fine adjustment which serves to place it in position near the centre of curvature. The screen is so close to the pin-hole in fact that both the source and a point on the edge of the screen may be said to be at the centre of curvature of the mirror. The mirror is temporarily mounted so as to have its axis horizontal, in a cellar or other place of uniform temperature.

The final focussing to the centre of curvature is made by the fine adjustment screws; the image may be received on a bit of paper placed on the screen and overlapping the edge nearest the source. The screws are worked till the image has its smallest dimension and is bisected by the edge of the screen. The test consists in observing the appearance of the mirror surface while the screen is tilted to cut off the light, as seen by an eye placed at the edge of the screen, a peephole or eye lens being provided to facilitate placing the eye in a correct position. The screen screws are worked so as to gradually cut off the light, and the observer notes the appearance of the mirror surface. If the curves are perfect and spherical, the transition from complete illumination to darkness will be abrupt, and no part of the mirror will remain illuminated after the rest.

For astronomical purposes a parabolic mirror is required. In this case the disc may be partially screened by zonal screens, and the position of the image for different zones noted; the correctness or otherwise of the curvature may then be ascertained by calculation.

A shorter way is to place the source just outside the focus, to be found by trial, and then, moving the extinction screen (now a separate appliance) to, say, five times the radius of curvature away, where the image should now appear, the suddenness of extinction may be investigated. This, of course, involves a corresponding modification of the apparatus.

Whether the tests indicate that a deepening of the Centre, i.e. increase of the curvature, or a flattening of the edges is required, at least two remedial processes are available. The "chisel and mallet" method of altering the size of the pitch, squares of the polisher may be employed, or paper or small pitch tools may be used to deepen the centre. The "chisel and mallet" method merely consists in removing pitch squares from a uniformly divided tool surface by means of the instruments mentioned. This removal is effected at those points at which the abrasion requires to be reduced.

When some practice is attained, I understand that it is usual to try for a parabolic form at once, as soon as the polishing commences. This is done by dividing the pitch surface by V-shaped grooves, the sides of the grooves being radii of the circular surface, so that the central parts of the mirror get most of the polishing action. If paper tools are used they must not be allowed much overhang, or the edges of the mirror betray the effects of paper elasticity. Most operators "sink" the middle, but the late Mr. Lassell, a most accomplished worker, always attained the parabolic form by reducing the curvature of the edges of a spherical mirror.

72. Preparation of Flat Surfaces. —

As Sir H. Grubb has pointed out, this operation only differs from those previously described in that an additional condition has to be satisfied. This condition refers to the mean curvature, which must be exact (in the case of flats it is of course zero) to a degree which is quite unnecessary in the manufacture of mirrors or lenses.

A little consideration will show that to get a surface flat the most straightforward method is to carry out the necessary and sufficient condition for three surfaces to fit each other impartially. If they each fit each other, they must clearly all be flat. To carry out the process of producing a flat surface, therefore, two tools are made, and the glass or speculum is ground first on one and then on the other, the tools being kept "in fit" by occasional mutual grinding. The grinding and polishing go on as usual. If paper is employed, care must be taken that the polisher is about the same size as the object to be polished.

There is a slight tendency to polish most at the edges; but if the sweeps are of the right shape and size, this may be corrected approximately. The best surfaces which have come under my notice are those prepared as "test surfaces" by Mr. Brashear of Alleghany, Pa., U.S.A. These I believe to be pitch polished. A pitch bed is prepared, I presume, in a manner similar to that described for rocksalt surfaces; but the working of the glass is an immense art, and one which I believe — if one may judge by results — is only known to Mr. Brashear.

In general, the effect of polishing will be to produce a convex or concave surface, quite good enough for most purposes, but distinctly faulty when tested by the interference fringes produced with the aid of the test plate. The following information therefore — which I draw from Mr. 'Cook — will not enable a student to emulate Mr. Brashear, but will undoubtedly help him to get a very much better surface than he usually buys at a high price, as exhibited on a spectroscope prism.

The only difference between this process and the one described for polishing lenses, lies in the fact that the rouge is put into the paper surface while the latter is wet with a dilute gum "mucilage." It is of course assumed that the object and the two tools have been finely ground and fit each other impartially. The paper is rubbed over with rouge and weak gum water. The tool, when dry, is applied to the flat ground surface (of the object), and is scraped with the three-cornered file chisel as formerly described. This process must be very carefully carried out. The paper must be of the quality mentioned, or may even be thinner and harder. The cross strokes should be more employed than in the case of the curved surfaces.

A good deal will depend on the method employed for supporting the work; it is in general better to support the tool, which may have a slate backing of any desired thickness, whereby the difficulty resulting from strains is reduced. The work must be mounted in such a way as to minimise the effect of changes of temperature. If a pitch bed is selected, Mr. Brashear's instructions for rock salt may be followed, with, of course, the obvious necessary modifications. See also next section.

73. Polishing Flat Surfaces on Glass or on Speculum Metal. —

The above process may be employed for speculum metal, or pitch may be used. In the latter case a fresh tool must be prepared every hour or so, because the metal begins to strip and leave bits on the polisher; this causes a certain amount of scratching to take place. As against this disadvantage, the process of polishing, in so far as the state of the surface is concerned, need not take an hour if the fine grinding has been well done.

For the finest work changes of temperature, as in the case of glass, cause a good deal of trouble, and the operator must try to arrange his method of holding the object so as to give rise to the least possible communication of heat from the hand.

The partial elasticity of paper, which is its defect as a polishing backing, is, I believe, partly counterbalanced by the difficulty of forming with pitch an exact counterpart tool without introducing a serious rise of temperature (i.e. warming the pitch). The rate of subsidence of the latter is very slow at temperatures where it is hard enough to work reliably as a polisher.

A student interested in the matter of flat surfaces will do well to read an account of Lord Rayleigh's work on the subject, Nature, vol. xlviii., 1893, pp. 212, 526 (or B. A. Reports, 1893). In the first of these communications Lord Rayleigh describes the method of using test plates, and shows how to obtain the interference fringes in the clearest manner.

For the ordinary optician a dark room and a soda flame afford all requisite information; and if a person succeeds in making three glass discs, say 6 inches in diameter, so flat that, when superposed in any manner, the interference fringes are parallel and equidistant, even to the roughest observation, he has nothing to learn from any book ever written on glass polishing. Lord Rayleigh has also shown how to use the free clean surface of water as a natural test plate.

Since the above was written the following details of his exact course of procedure have been sent to me by Mr. Brashear, and I hereby tender my thanks: —

"It really takes years to know just what to do when you reach that point where another touch either gives you the most perfect results attainable, or ruins the work you have already done. It has taken us a long time to find out how to make a flat surface, and when we were called upon to make the twenty-eight plane and parallel surfaces for the investigation of the value of the metre of the international standard, every one of which required an accuracy of one-twentieth of a wave length, we had a difficult task to perform. However, it was found that every surface had the desired accuracy, and some of them went far beyond it.

It is an advantage in making flat surfaces to make more than one at a time; it is better to make at least three, and in fact we always grind and 'fine' three together. In making speculum plates we get up ten or twelve at once on the lead lap. These speculum plates we can test as we go on by means of our test plane until we get them nearly flat. In polishing them we first make quite a hard polisher, forming it on a large test plane that is very nearly correct. We then polish a while on one surface and test it, then on a second and test it, and after a while we accumulate plates that are slightly concave and slightly convex. By working upon these alternately with the same polisher, we finally get our polisher into such shape that it approximates more and more to a flat surface, and with extreme care and slow procedure we finally attain the results desired.

All our flats are polished on a machine which has but little virtue in itself, unmixed with brains. Any machine giving a straight dia-metrical stroke will answer the purpose. The glass should be mounted so as to be perfectly free to move in every direction — that is to say, perfectly unconstrained. We mount all our flats on a piece of body Brussels carpet, so that every individual part of the woof acts as a yielding spring. The flats are held in place by wooden clamps at the edges, which never touch, but allow the bits of glass or metal to move slowly around if they are circular; if they are rectangular we allow them to tumble about as they please within the frame holding them.

For making speculum metal plates either plane or concave we use polishers so hard that they scratch the metal all over the surface with fine microscopic scratches. We always work for figure, and when we get a hard polisher that is in proper shape, we can do ever so many surfaces with it if the environments of temperature are all right. If we have fifty speculum flats to make, and we recently made three times that number, we get them all ready and of accurate surface with the hard polisher. Then we prepare a very soft polisher, easily indented when cold with the thumb nail. A drop of rouge and about three drops of water are put on the plate, and with the soft polisher about one minute suffices to clean up all the scratches and leave a beautiful black polish on the metal. This final touch is given by hand; if we do not get the polish in a few minutes the surface is generally ruined for shape, and we have to resort to the hard polisher again.

I assure you that nothing but patience and perseverance will master the difficulty that one has to encounter, but with these two elements 'you are bound to get there.'"

CHAPTER III

MISCELLANEOUS PROCESSES

74. Coating Glass with Aluminium and Soldering Aluminium. —

A process of coating glass with aluminium has been lately discovered, which, if I mistake not, may be of immense service in special cases where a strongly adherent deposit is required. My attention was first attracted to the matter by an article in the *Archives des Sciences physiques et naturelles de Gen趨*, 1894, by M. Margot. It appears that clean aluminium used as a pencil will leave a mark on clean damp glass. If, instead of a pencil, a small wheel of aluminium — say as big as a halfpenny and three times as thick — is rotated on the lathe, and a piece of glass pressed against it, the aluminium will form an adherent, though not very continuous coating on the glass.

Working with a disc of the size described rotating about as fast as for brass-turning, I covered about two square inches of glass surface in about five minutes. The deposit was of very uneven thickness, but was nearly all thick enough to be sensibly opaque. By burnishing the brilliance is improved (I used an agate burnisher and oil), but a little of the aluminium is rubbed off. The fact that the burnisher does not entirely remove it is a sign of the strength of the adherence which exists between the aluminium and the glass. In making the experiment, care must be taken to have the glass quite clean — or at all events free from grease — in order to obtain the best results.

M. Margot has contributed further information to the Archives des Sciences physiques et naturelles (February 1895). He finds that adherence between aluminium and glass is promoted by dusting the glass with powders, such as rouge. There is no doubt that a considerable improvement is effected in this way; both rouge and alumina have in my hands greatly increased the facility with which the aluminium is deposited. M. Margot finds that zinc and magnesium resemble aluminium in having properties of adherence to glass, and, what is more, carry this property into their alloys with tin. Thus an alloy of zinc and tin in the proportions of about 92 per cent tin and 8 per cent zinc may be melted on absolutely clean glass, and will adhere strongly to it if well rubbed by an asbestos crayon.

A happy inspiration was to try whether these alloys would, under similar circumstances, adhere to aluminium itself, and a trial showed that this was indeed the case, provided that both the aluminium and alloy are scrupulously clean and free from oxide. In this way M. Margot has solved the problem of soldering aluminium. I have satisfied myself by trial of the perfect ease and absolute success of this method. The alloy of zinc and tin in the proportions above mentioned is formed at the lowest possible temperature by melting the constituents together. It is then poured so as to form thin sticks.

The aluminium is carefully cleaned by rubbing with a cuttle bone, or fine sand, and strong warm potash. It is then washed in water and dried with a clean cloth. The aluminium is now held over a clean flame and heated till it will melt the solder which is rubbed against it. The solder sticks at once, especially if rubbed with another bit of aluminium (an aluminium soldering bit) similarly coated. To solder two bits of aluminium together it is only necessary to tin the bits by this process and then sweat them together.

The same process applies perfectly to aluminium caused to adhere to glass by the previously mentioned process, and enables strong soldered contacts to be made to glass. In one case, while I was testing the method, the adhesion was so strong that the solder on contracting while cooling actually chipped the surface clean off the glass. In order to get over this I have endeavoured to soften the solder by mixing in a little of the fusible metal mercury amalgam; and though this prevents the glass from being so much strained, it reduces the adherence of the solder. It is a comfort to be able to solder aluminium after working for so many years by way of electroplating, or filing under solder. An alternative method of soldering aluminium will be described when the electroplating of aluminium is discussed, 138.

Gilding Glass. — In looking over some volumes of the Journal fuer praktische Chemie, I came across a method of gilding glass due to Boettger (Journ. f. prakt. Chem. 103, p. 414). After many trials I believe I am in a position to give definite instructions as to the best way of carrying out this rather troublesome operation. The films of gold obtained by the process are very thick, and the appearance of the gold exceedingly fine. The difficulty lies in the exact apportionment of the reducing solution. If too much of the reducing solution be added, the gold deposits in a fine mud, and no coating is obtained. If, on the other hand, too little of the reducing solution be added, little or no gold is deposited. The secret of success turns on exactly hitting the proper proportions.

The reducing solution consists of a mixture of aldehyde and glucose, and the difficulty I have had in following Boettger's instructions arose from his specifying "commercial aldehyde" of a certain specific gravity which it was impossible to reproduce. I did not wish to specify pure aldehyde, which is not very easily got or stored, and consequently I have had to determine a criterion as to when the proportion of reducing solution is properly adjusted.

The aldehyde is best made as required. I employed the ordinary process as described in Thorpe's Dictionary of Applied Chemistry, by distilling alcohol, water, sulphuric acid, and manganese dioxide together. The crude product is mixed with a large quantity of calcium chloride (dry — not fused), and is rectified once. The process is stopped when the specific gravity of the product reaches 0.832 at 60 F. The specific gravity of pure aldehyde is 0.79 nearly.

The following is a modification of Boettger's formula:—

Solution I

1 gram of pure gold is converted into chloride — got acid free — i.e. to the state represented by $AuCl_3$, and dissolved in 120 cc. of water.

This solution is the equivalent of one containing 6.5 grains of trichloride to the ounce of water.

Solution II.

6 grams sodium hydrate.

100 grams water.

Solution III.

0.2 grams glucose (bought as pure).

12.6 cubic centimetres 95 per cent alcohol.

12.6 cubic centimetres water.

2.0 cubic centimetres aldehyde, sp. gr. 0.832.

To gild glass these solutions are used in the following proportions by volume:-

16 parts of No. I.

4 parts of No. II.

0.8 parts of No. III.

The glass is first cleaned well with acid and washed with water: it is then rinsed with Solution No. III. If it is desired to gild the inside of a glass vessel, Solution No. III. may be placed in the vessel first, and the walls of the vessel rinsed round carefully. Solutions I. and II. are mixed separately and then added to III. — after about two minutes the whole is well shaken up.

If it be desired to gild a mirror of glass, the glass-plate is suspended face downwards in a dish of the mixed solutions — care being taken to rinse the glass with Solution III. first.

If the mixture darkens in from 7' to 10' in diffuse daylight and at 60F. it will gild well, and it generally pays to make a few trials in a test tube to arrive at this. If too much reducing solution is present the liquid will get dark more rapidly, and vice versa. The gilding will require several hours — as much as twelve hours may be needed.

The reaction is one of great chemical interest, being one of that class of reactions which is greatly affected by capillarity. Thus it occasionally happens that when the reducing solution is not in the right proportion, gold will be deposited at the surface of the liquid (so as to form a gilt ring on the inside of a test tube), the remainder of the gold going down as mud. The gold deposited is at first transparent to transmitted light and is deeply blue. I thought this might be due to a trace of copper or silver, but on carefully purifying the gold no change of colour was noted. If the reducing solution is present in slightly greater proportions than that given in the formula, the gold comes down with a richer colour, and has a tendency to form a mat surface and to separate from the glass. The gold which is deposited more slowly has a less rich colour but a brighter surface. The operation should be interrupted when a sufficient deposit has been obtained, because it is found that the thicker the deposit, the more lightly is it held to the glass surface.

75. The Use of the Diamond-cutting Wheel. —

A matter which is not very well known outside geological circles is the manipulation of the diamond-cutting wheel, and as this is often of great use in the physical laboratory, the following notes may not be out of place. I first became acquainted with the art in connection with the necessity which arose for me to make galvanometer mirrors out of fused quartz, and it was then that I discovered with surprise how difficult it is to obtain information on the point. I desire to express my indebtedness to my colleagues, Professor David and Mr. Smeeth, for the instruction they have given me. In what follows I propose to describe their practice rather than my own, which has been of a makeshift description. I will therefore select the process of cutting a slice of rock for microscopical investigation.

76. Arming a Wheel. —

Fig.

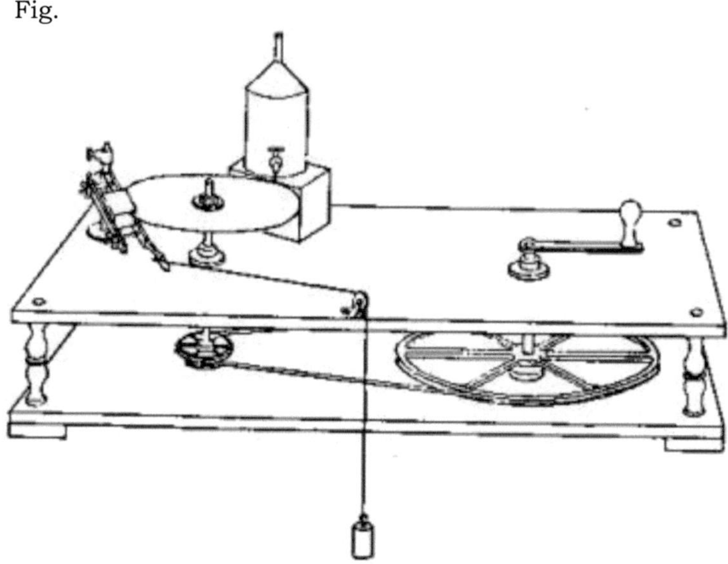

63.

A convenient wheel is made out of tin-plate, i.e. mild steel sheet, about one-thirtieth of an inch thick and seven inches in diameter. This wheel must be quite flat and true, as well as round; too much pains cannot be taken in securing these qualities. After the wheel is mounted, it is better to turn it quite true by means of a watchmaker's "graver", or other suitable tool. The general design of a rock-cutting machine will be clear from the illustration (Fig. 63).

The wheel being set up correctly, the next step is to arm it with diamond dust. For this purpose it is before all things necessary that real diamond dust should be obtained. The best plan is to procure a bit of "bort" which has been used in a diamond drill, and whose properties have therefore been tested to some extent. This is ground in a diamond mortar — or rather hammered in one — and passed through a sieve having at least 80 threads to the inch. The dust may be conveniently kept in oil.

To arm the wheel, a little dust and oil is taken on the finger, and laid on round the periphery of the wheel. A bit of flint or agate is then held firmly against the edge of the wheel and the latter is rotated two or three times by hand. The rotation must be quite slow — say one turn in half a minute — and the flint must be held firmly and steadily against the wheel. Some operators prefer to hammer the diamond dust into the wheel with a lump of flint, or agate, but there is a risk of deforming the wheel in the process. When a new wheel is set up, it may be necessary to repeat the above process once every half hour or so till the cutting is satisfactory, but when once a wheel is well armed it will work for a long time without further attention.

77. Cutting a Section. —

A wheel 7 inches in diameter may be rotated about 500 times per minute, and will give good results at that speed. The work, as will be seen from the diagram, is pressed against the edge of the wheel by a force, which in the case quoted was about the weight of eleven ounces. This was distributed along a cutting arc of three-quarters of an inch.

A convenient cutting lubricant is a solution of Castile soap in water, and this must be freely supplied; if the wheel gets dry it is almost immediately spoiled owing to the diamond dust being scraped off. In the figure the lubricant is supplied by a wick running into the reservoir. I have used both clock oil and ordinary gas-engine oil as lubricants, with equally satisfactory results. As to the speed of cutting, in the experiment quoted a bit of rather friable "gabbro," measuring three-quarters of an inch on the face by five-eighths of an inch thick, was cut clean through in six minutes, or by 3000 turns of the wheel. The travel of the edge was thus between 5000 and 6000 feet, or say 9000 feet, nearly 2 miles, per inch cut.

A good solid rock, like basalt, can be cut into slices of about 3/32 inch thick. A very loose rock is best boiled in Canada balsam, hard enough to set, before it is put against the wheel.

Instead of a grinding machine a lathe may be employed. The disc is, of course, mounted on the mandrel, and the work on the slide-rest. The latter must be disconnected from its feed screws, and a weight arranged over a pulley so as to keep the work pressed against the wheel by a constant force.

It may, perhaps, occur to the reader to inquire whether any clearance in the cut is necessary. The answer is that in all probability, and in spite of every care, the wheel will wobble enough to give clearance. If it does not, a little diamond dust rubbed into the side of the wheel, as well as the edge, will do all that is required. The edge also, after two or three armings, "burrs" a little, and thus provides a clearance naturally. It is not unlikely that in the near future the electric furnace will furnish us with a number of products capable of replacing the diamond as abrading agents. The cost of the small amount of diamond dust; required in a laboratory is so small, however, that it; is doubtful whether any appreciable economy will be, effected.

78. Grinding Rock Sections, or Thin Slips of any Hard Material.—

A note on this is, perhaps, worth making, for the same reasons as were given for note, 75, which it naturally follows. Just as trout-fishing; is described by Mr. Francis as the "art of fine and far off," *[Footnote:* In the Badminton Library, volume on Fishing.] section grinding may be called "the art of Canada balsam cooking," as follows. A section of rock having been cut from the lump as just described, it becomes; necessary to grind it down for purposes of microscopical investigation. For this purpose it is placed on a slip of glass, and cemented in position by Canada, balsam. Success in the operation of grinding the mounted section depends almost entirely on the way in which the mounting is done, and this in its turn depends on the condition to which the Canada has been brought.

To illustrate the operations, I will describe a specific case, viz. that of grinding the section of "gabbro"' above described, for microscopical purposes. One side of the section is probably sufficiently smooth and plane from the operation of the diamond wheel; if not, it must be ground by the finger on a slab of iron or gun-metal with emery and water, the emery passing a sieve of 80 threads to the inch. The glass base on which the section is to be mounted for grinding is placed on a bit of iron or copper plate over a Bunsen burner, and three or four drops of natural Canada balsam are placed upon it. The section is placed on the plate to heat at the same time.

The temperature must not rise so high as to cause any visible change in the Canada balsam, except a slight formation of bubbles, which rise to the surface, and can be blown off. The heating may require to be continued, say, up to twenty minutes. The progress of the operation is tested by examining the balsam as to its viscous properties.

An exceedingly simple and accurate way of testing is to dip a pair of ordinary forceps in the balsam, which may be stirred a little to secure uniformity. The forceps are introduced with the jaws in contact, and, as soon as withdrawn, the jaws are allowed to spring apart, thus drawing out a balsam thread. In a few moments the thread is cold, and if the forceps be compressed, this thread will bend.

The Canada must be heated until it is just in such a state that on bringing the jaws together the thread breaks. The forceps may open to about three-quarters of an inch. If the Canada is more viscous, so that the thread does not break, the section when cemented by it will most probably slip on the slide. On the other hand, if the balsam is more brittle, it will crumble away during the grinding.

Assuming that the proper point has been reached, the section is mounted with the usual precautions to avoid air bubbles, i.e. by dropping one edge on the balsam first. When all is cold, the surface of the section may be ground on an iron plate with emery passing the 80 sieve, till it is about 1/40 inch thick. From this point it must be reduced on ground glass by flours of emery and water; the rough particles of the former may be washed out for fine work.

The process of grinding should not take more than half an hour if the section is properly cut, etc. Beyond this point the allowable thickness must depend on the nature of the rock; a good general rule is to get the section just so thin that felspars show the yellow of the first order in a polarising, microscope. The section is then finished with, say, two minutes emery or water of Ayr-stone dust. It is better not to have the surface too smooth.

To transfer the section, the hard Canada round the sides is scraped away, and the section itself covered with some fresh Canada from the bottle. It is then warmed till it will slip off when a pin, or the invaluable dentist's chisel, is pressed against one side. If the section be very delicate, the cover slip should be placed over it before it is moved to the proper slide. The Canada used for mounting is not quite so hard as that employed in grinding, but it should be hard when cold, i.e. not sticky.

The art of preparing Canada balsam appears to consist in heating it under such conditions as will ensure its being exposed in thin layers. I have wasted a good deal of time in trying to bake Canada in evaporating basins, with the invariable result that it was either over or under-baked, and got dark in colour during the process.

On reviewing the process of rock section-cutting and mounting as just described, I cannot help thinking that, if properly systematised, it could be made much more rapid by the introduction of proper automatic grinding machinery. It also seems not improbable that a proper overhaul of available gums and cements would be found to lead to a cementing material less troublesome than Canada balsam.

79. Cutting Sections of Soft Substances. —

Though this art is fully treated of in books on practical biology, it is occasionally of use to the physicist, and the following note treats of that part of the subject which is not distinctly biological.

Soft materials, of which thin sections may be required, generally require to be strengthened before they are cut. For this purpose a variety of materials are available. The one most generally used is hard paraffin. The only point requiring attention is the embedding. The material must be dry.

This is accomplished by soaking in absolute alcohol, i.e. really absolute alcohol made by shaking up rectified spirit with potassium carbonate, previously dried, and then digesting for a day with large excess of quick-lime, making use of an inverted condenser and finally distilling off the alcohol without allowing it to come in contact with undried air. After soaking for some time in absolute alcohol, the material may be transferred to oil of bergamot, or oil of cloves, or almost any essential oil. After soaking in this long enough to allow the alcohol to diffuse out, the material may be lifted into a bath of melted paraffin (melting at, say, 51 C.). The process of soaking is in some cases made to go more rapidly by exhausting, and, if the material will stand it, by raising the temperature over 100 C. The soaking process may require minutes, hours, or days, according to the size and density of the material; but a few hours are usually sufficient.

When cold, the sections may be cut in any of the ordinary forms of microtome.

Another way of embedding is to soak in collodion, and then precipitate the latter in the material and around it by plunging into nearly absolute alcohol. The collodion yields a harder matrix than the paraffin.

Whatever form of cutting machine is employed, the art of sharpening the knife is the only one requiring any particular notice. The easiest way of obtaining a knife hard enough to sharpen, is to use a razor of good quality. If it has to be ground, it is best to do this on a fine Turkey stone which is conveniently rested on two bits of rubber tubing, to avoid jarring the blade. Many stones are slightly cracked, but on no account must the razor be dragged across a crack, or the edge will suffer.

The necessary and sufficient condition is that the razor must be worked in little sweeps over the stone, and pressed against the latter by little more than its own weight, and the grinding must be regular. The edge may be inspected under a microscope, and it must be perfectly smooth and even before it will cut sections. A finishing touch may be given on a leather strap, but it must be done skilfully, otherwise it is better omitted.

The necessity for providing exceptionally keen and sharp edges arose in the manufacture of phonographs, where the knife used to turn up the wax cylinders must leave a perfectly smooth surface. In

1889 this was being accomplished on an ivory lap fed with a trace of very fine diamond dust.

I have had this method in mind as a possible solution of the difficulty of razor-grinding, but have not tried it. I imagine one would use a soft steel or ivory slip rubbed over with fine diamond dust and oil by means of an agate. The lap used in the phonograph works was rotated at a high speed.

80. On the Production of Quartz Threads.' —

[Footnote: Since this was written an article on the same subject by Mr. Boys appeared in the Electrician for 1896. The instructions therein given are in accordance with what I had written, and I have made no alteration in the text.]

In 1887 the important properties of fused quartz were discovered by Mr. Vernon Boys (Philosophical Magazine, June 1887, p. 489, "On the Production, Properties, and Some Suggested Uses of the Finest Threads"). A detailed study of the properties of quartz threads was made by Mr. Boys and communicated to the Society of Arts in 1889 (Journal of the Society of Arts, 1889). An independent study of the subject was made by the present writer in 1889 (Philosophical Magazine, July 1890, "On the Elastic Constants of Quartz Threads ").

There is also a paper in the Philosophical Magazine for 1894 (vol. xxxvii. p. 463), by Mr. Boys, on "The Attachment of Quartz Fibres." This paper also appeared in the Journal of the Physical Society at about the same date, together with an interesting discussion of the matter. In the American Journal, Electric Power, for 1894, there is a series of articles by Professor Nichols on "Galvanometers," in which a particular method of producing quartz threads is recommended. The method was originally discovered by Mr. Boys, but he seems to have made no use of it. A hunt through French and German literature on the subject has disclosed nothing of interest — nothing indeed which cannot be found in the papers mentioned.

81. Quartz fibres have two great advantages over other forms of suspension when employed for any kind of torsion balance, from an ordinary more or less "astatic" galvanometer to the Cavendish apparatus. In the first place the actual strength of the fibres under longi-

tudinal stress is remarkably high, ranging from fifty to seventy tons weight per square inch of section, and even more than this in the case of very fine threads; the second and more important point in favour of quartz depends on the wide limits within which cylindrical threads of this material obey the simplest possible law of torsion, i.e. the law that for a given thread carrying a given weight at a given temperature and having one end clamped, the twist about the axis of figure produced by a turning moment applied at the free end is proportional simply to the moment of the twisting forces, and is independent of the previous history of the thread.

It is to be noted, however, that the torsional resilience of quartz as tested by the above law is not so perfect but that our instrumental means allow us to detect its imperfections, and thus to satisfy ourselves that threads made of quartz are not things standing apart from all other materials, except in the sense that the limits within which they may be twisted without deviating in their behaviour from the law of strict proportionality by more than some unassigned small quantity, are phenomenally wide.

A torsion balance — if we except the case of certain spiral springs — is almost always called upon for information as to the magnitude of very small forces, and for this purpose it is not essential merely that some law of twisting should be exactly obeyed, but also that the resistance to twisting of the suspension should be small.

Now, regarded merely as a substance possessing elastic rigidity, quartz is markedly inferior to the majority of materials, for it is very stiff indeed; its utility depends as much as anything upon its great strength, for this allows us to, use threads of exceeding fineness. In addition to this it must be possible, and moreover readily possible, to obtain threads of uniform section over a sufficient length, or the rate of twist per unit length of the thread will vary in practice from point to point, so that the limits of allowable twist averaged over the whole thread may not be exceeded, and yet they may be greatly overpassed at particular points of the thread.

It is interesting to note that in the case of quartz we not only have a means for readily producing very uniform cylindrical threads, but that the limits of allowable rate of twist are so wide that a small departure from uniformity of section produces much less inconvenience than in the case of any other known substance.

82. There are three methods generally in use for drawing quartz fibres, all depending on the fact that quartz when fused is so viscous that it may be drawn into threads of great length, without these threads breaking up into drops, or indeed without their showing any sign of doing so. The surface tension of the melted quartz must, however, be very considerable, as may be seen by examining the shape of a drop of the molten material, and this suffices to impress a rigidly cylindrical form upon the thread, the great viscosity apparently damping down all oscillation.

The first method is the one originally employed by Mr. Boys. A needle of quartz is melted somewhere in its length and is then drawn out rapidly by a light arrow, to which one end of the needle is attached, and which is projected from a kind of crossbow.

A modification of this method, which the writer has found of service when very thick threads are required, is to replace the bow and arrow by a kind of catapult.

The third method, which yields threads of almost unmanageable fineness, depends on the experimental fact that when a fine point of quartz is held in a high pressure oxygen gas blow-pipe flame, the friction of the flame gases suffices to overcome the tendency of the capillary forces to produce a spherical drop, and actually causes a fine thread to be projected outwards in the direction of the flame.

83. A preliminary operation to any method is the production of a stick of fused quartz. This is managed as follows. A rock crystal or quartz pebble is selected and examined. It must be perfectly white, transparent, and free from dirt. Surface impurity can of course be got rid of by means of a grindstone. The crystal is placed in a perfectly clean Stourbridge clay crucible, furnished with a cover, and heated to bright redness for about an hour in a clean fire or in a Fletcher's gas furnace. The contents of the crucible are turned out when sufficiently cool on to a clean brick or bit of slate. It will be found that the crystal is completely broken up and the fragments must be examined in case any of them have become contaminated by the crucible, but this will not have happened if the temperature did not rise beyond a bright red heat.

The heap of fragments being found satisfactory, the next thing is to fuse some of the pieces together. Unless the preliminary heating has been efficiently carried out this will prove an annoying task, because a rock crystal generally contains so much water that it splinters under the blow-pipe in a very persistent manner. There are two ways of assembling the fragments. One is to place two tiles or bricks on edge about the heap of quartz lying upon a third tile, so that the heap occupies the angular corner or nook formed by the tiles (Fig. 64).

The oxygas blow-pipe previously described is adjusted to give its hottest flame, the bags being weighted by at least two hundred-weight, if of the size described (see 15).

The tip of the inner cone of the blow-pipe is brought to bear directly upon one of the fragments, and if the operation is performed boldly it will be found that the surface of the fragment can be fused, and the fragment thus caused to hold together before the lower side gets hot enough to suffer any contamination from the tile or brick. A second fragment may be treated in the same way, and then a third, and so on.

Finally, the fragments may be fused together slightly at the corners, and a stick may thus be formed. Of course a good deal of cracking and splitting of the fragments takes place in the process; the best pieces to operate upon are those which are well cracked to begin with, and that in such a way that the little fragments are interlocked.

An alternative method which has some advantages is to arm a pair of forceps with two stout platinum jaws, say an inch and a half long, and flattened a little at the ends. The fragments are held in these platinum forceps and the blow-pipe applied as before. This method works very well in adding to a rod which has already been partly formed, but the jaws require constant renewals. The first fragment which is fused sufficiently to cohere may also be fused to a bit of tobacco pipe, or hard glass tube or rod, and the quartz stick gradually built up by fusing fresh pieces on to the one already in position.

Fig.

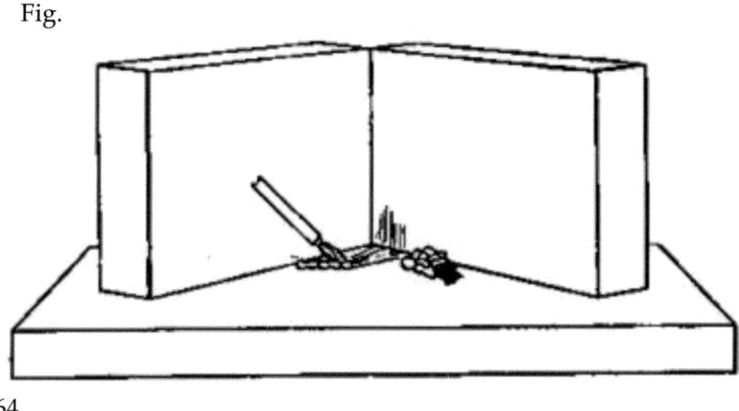

64.

Since the glass or pipeclay will contaminate the quartz which has been fused on to it, it is necessary to discard the end pieces at the close of the operation. A string of fragments having been collected and stuck together, the next step is to fuse them down into a uniform rod. This is easily done by holding the string in the blow-pipe flame and allowing it to fuse down. Twisting the fused part has a good effect in assisting the operation. It is desirable to use a large jet and as powerful a flame as can be obtained during this part of the operation.

The final result should be a rod, say two or three inches long and one-eighth of an inch thick, which will in most cases contain a large number of air bubbles. Since the presence of drawn-out bubbles cannot be advantageous, it is often desirable to get rid of them, and this can conveniently be done at the present stage. The process at best is rather tedious; it consists in drawing the quartz down very fine before an intense flame, in order to allow the bubbles to get close enough to the surface to burst. A considerable loss of material invariably occurs during the process; for whenever the thin rod separates into two bits the process of flame-drawing of threads goes on, and entails a certain waste; moreover, the quartz in fine filaments is probably partially volatilised.

Sooner or later, however, a sufficient length of bubble-free quartz can be obtained. It must not be supposed that it is always necessary to eliminate bubbles as perfectly as is contemplated in the foregoing description of the treatment, but for special purposes it may be essential to do so, and in any case the reader's attention is directed to a possible source of error.

It may be mentioned in connection with this matter that crystals of quartz may look perfectly white and clear, and yet contain impurity. For instance, traces of sodium are generally present, and lithium was found in large spectroscopic quantity in five out of six samples of the purest crystals in my laboratory. The presence of lithium in rock crystal has also been detected by Tegetmeier (Vied. Ann., xli. p. 19, 1890).

After some practice in preparing rods and freeing them of bubbles the operator will notice a distinct difference in the fusibility of the samples of quartz he investigates, though all may appear equally pure to the unaided eye. It should be mentioned, however, that high infusibility cannot always be taken as a test of purity, for the most infusible, or rather most viscous, sample examined by the writer contained more lithium than some less viscous samples.

Fig.  65.

During the process of freeing the quartz from bubbles the lithium and sodium will be found to burn away, or at all events to disappear.

A rod of quartz, say three inches long, one-sixteenth of an inch in diameter, and free from bubbles for half an inch of its length, even when examined by a strong lens, is suitable for drawing into threads. The rod is manipulated exactly in the manner described under GLASS-BLOWING, and is finally drawn down at the bubble free part into a needle, say 0.02 inch in diameter (No. 25 on the Birmingham wire gauge), and 2 inches long.

Fig.
66.

There is one peculiarity about fused quartz which renders its manipulation easier than that of glass — it is impossible to break fused quartz, however suddenly it be thrust into the blow-pipe flame. A rod having a diameter of three-sixteenths of an inch — and perhaps much more — may be brought right up to the tip of the inner cone

180

of the oxy-gas flame and held there-till one side fuses, the other being comparatively cool, without the slightest fear of precipitating a smash. In seven years' experience I have never seen a bit of once fused quartz broken by sudden heating; whether it might be done if sufficient precautions were taken I do not know.

The reason of the fortunate peculiarity of quartz in this respect is, I presume, to be found in the fact that quartz once it has been fused is really a very strong material indeed, and is also probably the least expansible substance known. From some experiments of the writer upon the subject, it may be concluded that at the most quartz which has been fused expands only about one-fifth as fast as flint-glass, at all events between 20 and 70 C.

84. Drawing Quartz Threads. —

The thick end of the rod of quartz is held in the fingers or occasionally in a clip. The end of the fine point is attached to a straw arrow by means of a little sealing-wax. The arrow is laid on the stock of a crossbow in the proper position for firing. See Figs. 67 and 68, which practically explain themselves.

The needle is heated by the blow-pipe till a minute length is in a state of uniform fusion; the arrow is then let fly, when it draws a thread out with it. The arrow is preferably allowed to strike a wooden target placed, say, 30 feet away from the bow, and a width of black glazed calico is laid under the line of fire to catch the thread or arrow if it falls short. The general arrangements will be obvious from the figure.

The bow is of pine in the case where very long thin threads are required, though for ordinary purposes I have found a bow of lance-wood succeed quite as well. The trigger of the bow consists of a simple pin passing through the stock and fastened at its lower end to a string connected with a board which can be depressed by foot. In the figure an ordinary trigger is shown, but the pin does just as well.

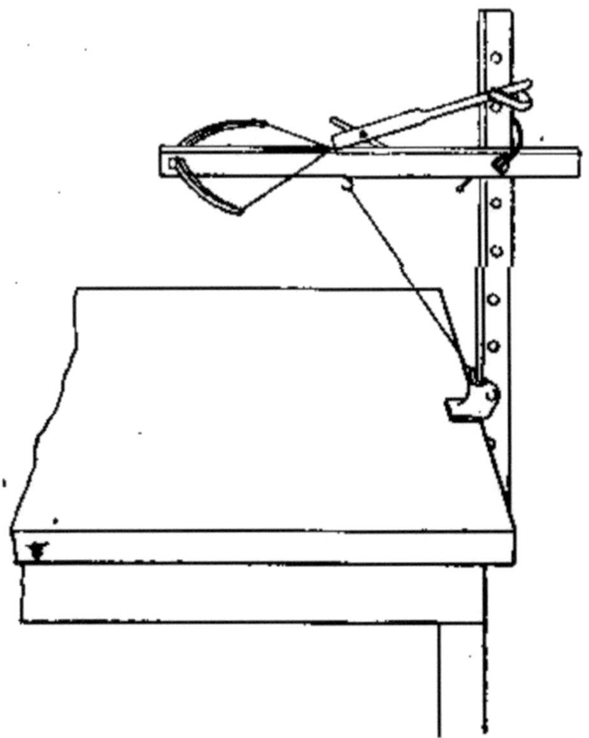

Fig 67

The arrow is made out of about 6 inches of straw, plugged up aft by a small plug of pine or willow fastened in with sealing-wax, and projecting backwards one-eighth of an inch. This projection serves a double purpose: it gives a point of attachment for the quartz needle, and on firing the bow it forms a resisting anvil on which the string of the bow impinges. The head of the arrow is formed by a large needle stuck in with sealing-wax, and heavy enough to bring the centre of gravity of the arrow forward of one-third of its length, the condition of stability in flight.

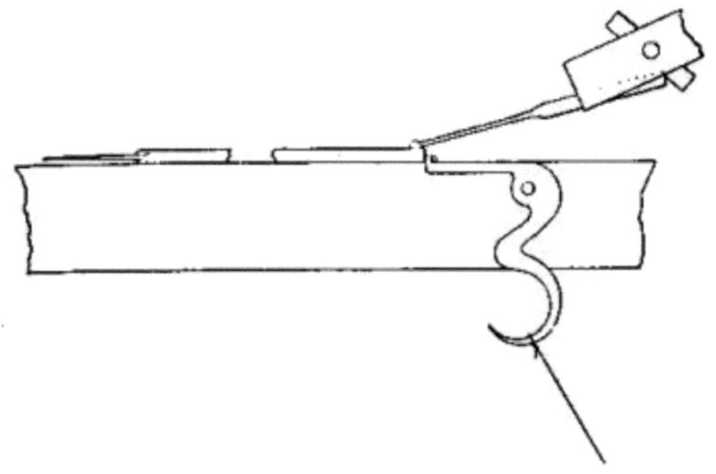

It is not necessary to employ any feathering for these arrows; though I have occasionally used feathers or mica to "wing the shaft" no advantage has resulted therefrom.

To get fine threads a high velocity is essential. This is obtained by considering (and acting upon) the principles involved. The bow may be regarded as a doubly-tapering rod clamped at the middle. After deflection it returns towards its equilibrium position at a rate depending in general terms on the elastic forces brought into play, directly, and on the effective moment of inertia of the rod, inversely (see Rayleigh, Sound, vol. ii. chap. viii.) If the mass of the arrow is negligible compared with the bow, the rate at which the arrow moves is practically determined by that attained by the end of the bow, which is a maximum in crossing its equilibrium position.

The extent to which the arrow profits by this velocity depends on the way the bow is strung. It will be greatest when the string is perpendicular to the bow when passing its equilibrium position; or in other words, when the string is infinitely long. Since the string has mass, however, it is not permissible to make it too long, or its weight begins to make itself felt, and a point is soon reached at which the geometrical gain in string velocity is compensated for by the total loss of velocity due to the inertia of the string. In practice it is sufficient to use a string 10 per cent longer than the bow.

It is well to use a light fiddle string, served with waxed silk at the trigger catch; if this be omitted the gut gets worn through very quickly. In order to decide how far it is permissible to bend the bow, the quickest way is to make a rough experiment on a bit of the same plank from which the bow is to be cut, and then to allow a small factor of safety. In the figure the bow is of lance-wood and is more bent than would be suitable for pine.

The bow itself is tapered from the middle outwards just like any other bow. If thick threads are required, the above considerations are modified by the fact that quartz opposes a considerable resistance to drawing, and that consequently the arrow must not only have a high velocity, but a fair supply of energy as well; in other words, it must be heavy. A thin pine arrow instead of a straw generally does very well, but in this case the advantage of using pine for the bow vanishes; and in fact lance-wood does better, owing to the greater displacement which it will stand without breaking. This of course only means that a greater store of energy can be accumulated at one bending.

I had occasion to investigate whether the unavoidable spin of an arrow about its axis produces any effect on the thread, and for this purpose made arrows with inertia bars thrust through the head, i.e. an arrow with a bit of wire run through it, perpendicular to its length — forming a cross in fact — the arms of the cross being weighted at the extreme ends by shot. This form of arrow has a considerable moment of inertia about its longer axis, and consequently rotates less than a mere straw, provided that the couples tending to produce rotation are not increased by the cross arm, or the velocity too much reduced. Shooting one of these arrows slowly, I could see that it did not rotate, and when fired at a high velocity, it generally arrived at the target (placed at varying distances front bow) with the arms nearly horizontal, thus showing that it probably did not rotate much.

I did not succeed in this at the first trial, by any means. The threads got in this way were no better than those made with a single straw, whence we may conclude very provisionally that the spin of the arrow has only a small effect, if any, on the quality of the threads.

Feathering the arrow, in my experience, tends, if anything to make it spin more; for one thing, because it is practically impossible to lay the feathering on straight.

After the arrow is shot, it remains to gather in the thread, and if the latter is at all thin, we have a rather troublesome job. In a thread thirty or forty feet long, the most uniform part generally lies in the middle if the thread is thin, i.e. of the order of a ten-thousandth of an inch in diameter. If the thread is thick the most uniform part may be anywhere. The part of the thread required is generally best isolated by passing a slip of paper under it at each end and cementing the thread to the paper by means of a little paraffin or soft wax, and then cutting off the outer portions. One bit of paper may then be lifted off the calico, and the thread will carry the other bit. In this way the thread may be taken to a blackened board, where it may be mounted for stock.

By passing the two ends of the thread under a microscope, or rather by breaking bits off the two ends and examining them together, it is easy to form an Opinion as to uniformity.

Mr. Boys has employed an optical method of examining threads, but the writer has invariably found a high-power microscope more convenient and capable of giving more exact information as to the diameter of the threads.

The beginner — or indeed the practised hand — need not expect to get a thread of the exact dimensions required at the first shot. A little experience is necessary to enable one to judge of the right thickness of the needle for a thread of given diameter. The threads are so easily shot, however, that a few trials take up very little time and generally afford quite sufficient experience to enable a thread of any required diameter to be prepared.

It is no use attempting to heat an appreciable length of needle; if this be done the thread almost invariably has a thick part about the middle of its length.. It is sufficient to fuse at most about one-twentieth of an inch along the needle before firing off the bow. This can be done by means of the smaller oxygas blow-pipe jet described in the article on blow-pipes for GLASS-BLOWING, 14. The flame must of course be turned down so as to be of a suitable size. A sufficiently small flame may be got from almost any jet.

If the needle be not equally heated all round, the thread tends to be curly; indeed by means of the catapult, threads may be pulled which, when broken, tend to coil up like the balance-springs of watches, if only care be taken to have one side of the needle much hotter than the other.

85. When examining bits of threads, say thicker than the two-thousandth of an inch, under the microscope it is convenient to use a film of glycerine stained with some kind of dye, in order to render the thread more sharply visible. The thread is mounted beneath a cover slip, and a drop of the stained glycerine allowed to run in. Such a treatment gives the image of the thread a sharply defined edge 3 and the contrast between the whiteness of the thread and the colour of the background allows measurements to be made with great ease.

On the whole the easiest way of measuring the diameter of a thick thread is to use a measuring microscope, i.e. one in which the lens system can be displaced along a plane bed by means of a finely cut micrometer screw. The instruments made by the Cambridge Scientific Instrument Company do fairly well. Direct measurements up to 0.0001 inch are easily made by means of a microscope provided with a Zeiss "A" objective, and rather smaller differences of thickness can be made out by it. For thin threads the method next to be described is more fitting, because higher powers can be more conveniently used.

In this method an ordinary microscope is employed together with a scale micrometer, and either an eyepiece micrometer, or a camera and subsidiary scale. The eyepiece micrometer is the more convenient. If a camera be employed, i.e. such an one as is supplied by Zeiss, it is astonishing how the accuracy of observation may be increased by attending carefully to the illumination of both the subsidiary scale and of the thread. The two images should be as far as possible of equal brightness, and for this purpose it will be found requisite to employ small screens.

The detail of making a measurement by means of the micrometer eyepiece is very simple. The thread is arranged on the stage so as to point towards the observer, and the apparent diameter is read off on the eyepiece scale. In order to calibrate the latter it is only necessary to replace the thread by the stage micrometer, and to observe the number of stage micrometer divisions occupying the space in the eyepiece micrometer formerly occupied by the thread. It is essential that both thread and stage micrometer should occupy the same position in the field, for errors due to unequal distortion may otherwise become of importance. For this reason it is best to utilise the centre of the field only.

The same remark applies to measurements by means of the camera, where the image of the thread is projected against the reflected image of the subsidiary scale laid alongside the microscope. In this case the value of the subsidiary scale divisions must be obtained from the divisions of the stage micrometer, coinciding as nearly as possible with the position occupied by the thread. Before commencing a measurement the screens are moved about till both images appear equally bright.

Threads up to about one twenty-thousandth of an inch in diameter may be sufficiently well measured by means of a Zeiss "4 centimetre apochromatic object-glass" and an eyepiece "No. 6" with sixteen centimetre tube length. *[Footnote: The objective certainly had "4 cm." marked on it, but the focal length appeared to be about I.5 mm. only.]*

86. Drawing Threads by the Catapult.

—

The bow-and-arrow method fails when threads of a greater diameter than about 0.0015 inch are required — at least if any reasonable uniformity be demanded, and no radical change in the bow and arrow be carried out.

Thus in the writer's laboratory a thread of about this diameter, within 1/10000 of an inch-13 inches long and free from air bubbles — was required. A fortnight's work by a most skilful operator only resulted in the production of two lengths satisfying the conditions.

The greatest loss of time occurs in the examination of the thread by means of the microscope.

Threads for galvanometer suspensions are conveniently from 0.0001 to 0.0004 inch in diameter, and are much more easily made and got uniform than thicker threads, to the production of which the catapult method applies.

A reference to the diagram will make the construction of the instrument quite clear. The moving end of the quartz is attached to a small boxwood slider working on a tubular girder or between wires. The quartz is secured in position by clamps shown at A and B, and motion is imparted to the slider by a stretched piece of catapult elastic (C). An easy means of regulating the pull of the elastic is to hold it back by a loop of string whose length can be varied by twisting it round a pin.

Fig.   69.

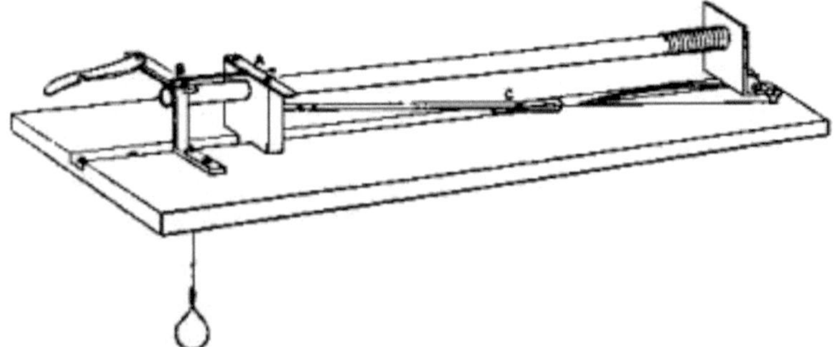

[Footnote: For greater clearness of drawing, the tube carrying the slider is shown somewhat higher above the base than is convenient in practice; and the slide itself is shown too thin in the direction of the hole through it.]

Since it is not permissible to allow the slider to rebound at the end of its journey, some such arrangement of breaks as is shown must be adopted. In the diagram the bottom of the slider runs on to a brass spring between the girder and the base of the appliance, and so gets jammed; the spiral spring acts merely as an additional guard. The diagram does not show the lower spring very clearly; it is a mere strip lying in the groove.

A rod of quartz, with a needle at one end, is prepared as before and secured in the clamps. During the operation of fastening down the clamps, there is some danger of breaking the needle, and consequently it is advisable to soften the latter before and while adjusting the second clamp.

The process of drawing a thread by this method is exactly similar to the operation already described in connection with the arrow method. Though short thick threads form the product generally obtained from the catapult, it must not be supposed that thin threads cannot be obtained in this way. If a short length of a very fine needle be heated, it will be found to yield threads quite fine enough for ordinary suspension purposes, but naturally not so uniform as those obtained from the 40-foot lengths obtainable by the bow-and-arrow method.

It is easy to make spiral quartz springs resembling watch balance-springs by means of the catapult. All that is necessary is to see that the quartz is rather unequally heated before the shot is fired. In the future it is by no means impossible that such springs may have a real value, for the rigidity of quartz is known to increase as temperature rises. Hence it is probable that the springs would become stiffer as temperature rises, even though they work chiefly by bending, and little or not at all by twisting. As this is the kind of temperature variation required to compensate an uncompensated watch balance wheel, it may turn out to have some value.

87. Drawing Threads by the Flame alone. —

A stick of quartz is drawn down to a fine point, and the tip of this point is held in the blow-pipe flame in the position shown in Fig. 70.

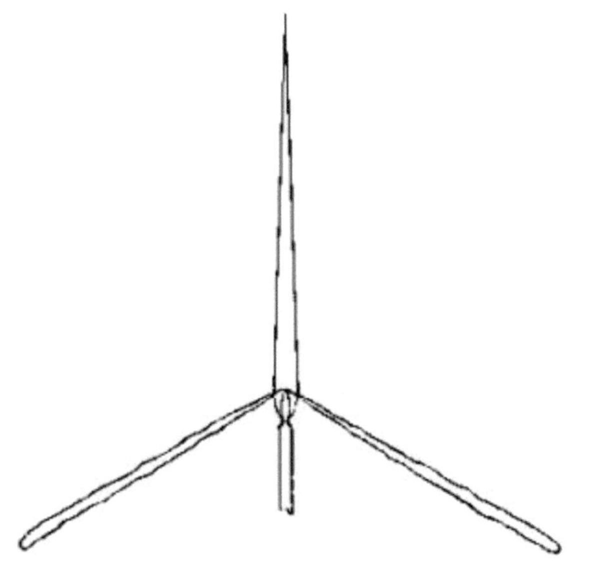

Fig. 70.

The friction of the flame gases is found to be sufficient to carry forward the fused quartz and to draw it into threads in spite of the influence of the capillary forces. If a sheet of paper be suspended at a distance of two or three feet in front of the blow-pipe flame, it will be found to be covered with fine threads tangled together into a cobwebby mass. As this method is an exceedingly simple one of obtaining threads, I have endeavoured to reduce it to a systematic operation.

A sheet of cardboard, about two feet square, is painted dead black and suspended horizontally, painted side downwards (Fig. 70, A), at a height of about two feet above the blow-pipe flame. The latter is adjusted so as to point almost vertically upwards and towards the centre of the cardboard. A few half-inch pins are thrust through the card from the upper surface and pushed home; about one dozen pins scattered over the surface will be sufficient. Their object is to prevent the threads being carried away round the edge of the screen.

The flame from the jet described so often is fed from gas bags weighted to about eighty pounds per square foot of (one) surface, i.e. "4-foot" bags require from three to four hundredweight to give an advantageous pressure. *[Footnote:* The resulting threads were really too fine for convenient manipulation, so that unless extremely fine threads are required it will be better to reduce the pressure of the gases considerably.*]*

Two sticks of quartz are introduced and caused to meet just in front of the inner cone — the hottest part of the flame. They are then drawn apart so as to form a fine neck, which softens and is bent in the direction of motion of the flame gases. When fusion is complete the neck separates into two parts, and a thread is drawn from each of them. By alternately lightly touching the rods together, and drawing them apart, quite a mass of threads may be obtained in two or three minutes, when the process should be stopped. If too many threads get entangled in the pins, one gives one's self the unnecessary trouble of separating them. On taking down the card it will be found that the threads have been caught by the pins; but the card now being laid black side upwards, the former easily slip off the points.

Threads at least a foot long, and perhaps vastly longer, may be obtained by this method, and are extraordinarily fine. When I first read Professor Nichols' statement (Electric Power, 1894) as to the value of these fibres for galvanometer purposes, I was rather sceptical on the ground that the threads would tend to get annealed by being drawn gradually, instead of suddenly, from a place of intense heat to regions of lower temperature.

Now annealing threads by a Bunsen makes them rotten. The threads being immersed in the hot flame gases could only cool at the same rate as the gas, and it was not — and is not — clear to me that annealing of the threads can be avoided. On the other hand, it may be possible that a thread cooled slowly from the first does not suffer in the same way as a cold thread would do when annealed in a Bunsen flame.

Again the velocity of the gases is beyond doubt exceedingly high, so that the annealing, even supposing it to be deleterious, might not be carried very far. Threads drawn by this method and measured "dry," i.e. by mounting them on a slide without the addition of any liquid, turned out to have a diameter of about 1/20000 of an inch.

I do not think I could manage to mount such fine threads without very special trouble. All the threads lying on the board, however, were found in reality to consist of three or four separate threads, and there is no reason why several threads should not be mounted in parallel, provided, of course, that they are equally stretched and touching each other. Equality of tension in the mounting could be secured by making one attachment good, then cementing the other attachment to the other end of the threads, and "drawing" the two attachments slightly apart at the moment the cement commences to set. This method may turn out to be very valuable, for, so far as I can see, the carrying power would be increased without an increase of torsional stiffness of anything like so high an order as would be the case were one thread only employed. On the other hand, the law of torsion could hardly be quite so simple, at all events, to the second order of approximations.

88. Properties of Threads. —

A large number of experiments on the numerical values of the elastic constants of quartz threads have been made by Mr. Boys and his students, and by the writer. As the methods employed were quite distinct and the results wholly independent, and yet in good agreement with each other, a rounded average may be accepted with considerable confidence.

TENACITY OF QUARTZ FIBRES (BOYS).

Diameter of Thread.		Tenacity in Tons' Weight per Square Inch of Section.	Tenacity in Dynes per Square Centimetre.
Inches	Centimetres		
0.000 69	0.00175	51.7	8×10^9
0.000 19	0.00048	74.5	11.5×10^9

Rounded mean of Boys' and Threlfall's results:

Young's Modulus at 20 C.,	5.6×10^{11} C.G.S.
Modulus of Simple Rigidity at 20 C.,	2.65×10^{11} C.G.S.
Modulus of Incompressibility,	1.4×10^{11} C.G.S.
Modulus of Torsion,	3.7×10^{11} C.G.S.

Approximate coefficient of linear expansion of quartz per degree between 80 C. and 30 C. is 0.0000017 (*Threlfall = loc. cit.*).

This must be regarded with some suspicion, as the data were not concordant. There is no doubt, however, about the extreme inexpansibility of quartz.

Temperature coefficient of modulus of torsional rigidity per degree centigrade, 22 to 98 C., 0.000133

Ditto, absolute simple rigidity, 0.000128 (*Threlfall*).

Limit of allowable rate of twist in round numbers is, one-third turn per centimetre, in a fibre 0.01 cm. diameter.

The limiting rate is probably roughly inversely as the diameter.

Attention must be called to the rapid increase in the torsional rigidity of these threads as the temperature rises. A quartz spiral spring-balance will be appreciably stronger in hot weather.

89. In the majority of instances in which quartz threads are applied in the laboratory, it is desirable to keep the coefficient of torsion as small as possible, and hence threads are used as fine as possible.

It is convenient to remember that a thread 0.0014 cm. or 0.0007 inch in diameter breaks with a weight of about ten grammes, and may conveniently be employed to carry, say, five grammes. With threads three times finer the breaking strength per unit area increases, say, 50 per cent. In ordinary practice — galvanometric work for instance — where it is desirable to use a thread as fine and short as possible to sustain a weight up to, say, half a gramme, it will be found that fibres five centimetres long or over give no trouble through defect of elastic properties. A factor of safety of two is a fair allowance when loading threads.

No difficulty will be experienced in mounting threads having a diameter of 0.0002 inch or over. With finer threads it is necessary to employ very dark backgrounds (Mr. Boys uses the darkness of a slightly opened drawer), or the threads cannot be sufficiently well seen.

In the case of instruments in which threads remain highly twisted for long periods of time, the above rule as to the safe limit of twist does not allow of a sufficient margin; it is only applicable to galvanometric and similar purposes.

The cause of the increase in tenacity as the diameter diminishes is at present unknown. It is due neither to an effect of annealing (annealed threads are rotten), nor is it a skin effect, nor is it due to the cooling of the thread under higher capillary pressure. It is, however, possible that it may be associated with some kind of permanent set taken by the fibres during the stage of passage from the liquid to the solid state.

90. On the Attachment of Quartz Fibres. —

For many purposes it is sufficient to cement the fibres in position by means of ordinary yellow shellac, but where very great accuracy is aimed at, the shellac (being itself imperfectly elastic and exposed to shearing stress) imposes its imperfections on the whole system. This source of error can be got over by soldering the threads in position. Attempts were made by the writer in this direction, with fair success, in 1889, but as Mr. Boys has carried the art to a high degree of perfection, I will suppress the description of my own method and describe his in preference. It has, of course, been frequently repeated in my laboratory.

In many cases, however, if not in all, it may be replaced by Margot soldering, as already described, a note on the application of which to this purpose will follow.

A thread of the proper diameter having been selected, it is cut to the right length. With fine threads this is not always a perfectly easy matter. The best way is for the operator to station himself facing a good light, not sunlight, which is too tiring to the eye, but bright diffused light. The thread will be furnished with bits of paper stuck on with paraffin at both ends, as already described.

A rough sketch of the apparatus — or, at all events, two lines showing the exact length which the free part of the thread must have — are marked on a smooth board, and this is supported with its plane vertical. The thread is held against the board, and the upper piece of paper is stuck lightly to the board with a trace of soft wax, so that the lower edge of the paper is at any desired height above the upper mark. This distance is measured, and forms the length of thread allowed to overlap the support. A second bit of paper is attached below the lower mark, a margin for the attachment of the lower end being measured and left as before. The thread will be most easily seen if the board is painted a dead black.

If it is desired to attach the thread to its supports merely by shellac, this is practically all that needs to be done. The supports should resemble large pins. The upper support will be a brass wire in most cases, and will require to be filed away as shown in the sketch (Fig. 71). It is then coated with shellac by heating and rubbing upon the shellac. As previously noted, the shellac must not be overheated.

The thread is cut off below the lower slip of paper, and the upper support being conveniently laid in a horizontal position on another dead-black surface, the thread is carried to it and laid as designed against the shellac, which is now cold. When the thread is in place, a soldering iron is put against the brass wire, and the shellac gradually melted till it closes over the thread.

Fig. 71.

The iron is then withdrawn and the thread pulled away from the point for one-twentieth of an inch or less. This ensures that the thread makes proper contact with the cement, and also that it is free from kinks; of course, it must leave the cement in the proper direction. A similar process is next carried out with respect to the lower attachment, and the ends of the thread are neatly trimmed off.

Both ends of the thread being secured, the next step is to transfer the upper support to a clip stand, the suspended parts being held by hand, so that the weight comes on the thread very gradually. In this way it will be easily seen whether the thread is bent where it enters the shellac, and should this be the case, a hot iron must be brought up to the shellac and the error rectified.

When both the support and the suspended parts are brought nearly to the required bearing, the hot iron is held for a moment close up to each attachment, the hand being held close below but not touching the suspended parts, and both attachments are allowed to straighten themselves out naturally.

These details may appear tiresome, and so they are when written out at length, but the time occupied in carrying them out is very short, and quartz threads break easily, unless the pull upon them is accurately in the direction of their length at all points.

In the event of its being decided to attach the thread by soldering, the process is rather more expensive in time, but not otherwise more troublesome.

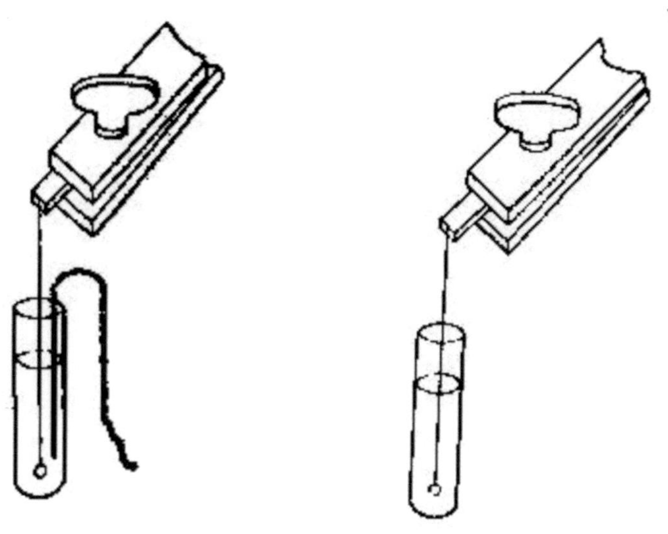

Fig. 73.

The thread being cut as before to the proper length, little bits of aluminium foil are smeared all over with melted shellac and suspended from the thread replacing the paper slips before described. It is important that no paraffin should be allowed to touch the thread anywhere near a point intended to be soldered. The thread is hung up from a clip stand by one of the bits of foil, and the lower end is washed by dipping it into strong nitric acid for a moment and thence into water. The object of smearing the foils all over with shellac is to prevent them being acted upon by the acid. The threads are not very easily washed acid free, but the process may be assisted by means of a fine camel's-hair pencil.

Some silvering solution made as described (65) is put into a test tube; the thread, after rinsing with distilled water, is lowered into the solution so far as is required, and is allowed to receive a coating of silver. It has been observed that the coating of silver must not be too thick — not sufficiently thick to be opaque. A watch may be kept on the process by immersing a minute strip of mica alongside the thread.

The silvered thread is rinsed with distilled water and allowed to dry.

Meanwhile the other end of the thread may be silvered. When both ends are silvered the process of coppering by electro deposit is commenced. A test tube is partially filled with a ten per cent solution of sulphate of copper, and several copper wires are dipped into it to form an anode. The thread is lowered carefully into the solution so as not to introduce air bubbles, and the silvered part is allowed to project far enough above the surface of the solution to come in contact with a fine copper wire. The circuit is closed through a Leclanché cell and a resistance box.

It is as well to begin with a fair resistance, say 100 ohms out in the box, and the progress of the deposit is watched by means of a low-power microscope set up in front of the thread. If the copper appears to come down in a granular form, the resistance is too small and must be increased; if no headway appears to be made, the resistance must be diminished.

As soon as a fair coat of copper has come down, i.e. when the diameter of the thread is about doubled, the process is interrupted. The thread is withdrawn, washed, dipped in a solution of chloride of zinc, and carefully tinned by dragging it over a small clean drop of solder on a soldering bit.

During this part of the process the shellac is apt to get melted if the iron is held too close, so that it is advisable to begin by making the thread somewhat over long. The end of the thread must only be trimmed off at the conclusion of the operation, i.e. after the thread is soldered up. The thread is attached to the previously tinned supports much in the same way as has been described under the head of shellac attachments. It does not very much matter whether both ends are coppered before one is soldered up or not. At the conclusion of the whole process the superfluous copper and silver are dissolved off by a little hot strong nitric acid applied on a glass hair pencil. This is best done by holding the thread horizontally with the assistance of clip stands.

If the thread is too delicate to bear brushing, the nitric acid may be applied by pouring out a big drop into a bit of platinum foil and holding this below the thread so as to touch it lightly. The dissolving of the copper and silver is, of course, followed by copious washing with hot water. This process is more laborious than might be imagined, but it may be shortened by heating the platinum foil supporting the water (Fig. 74).

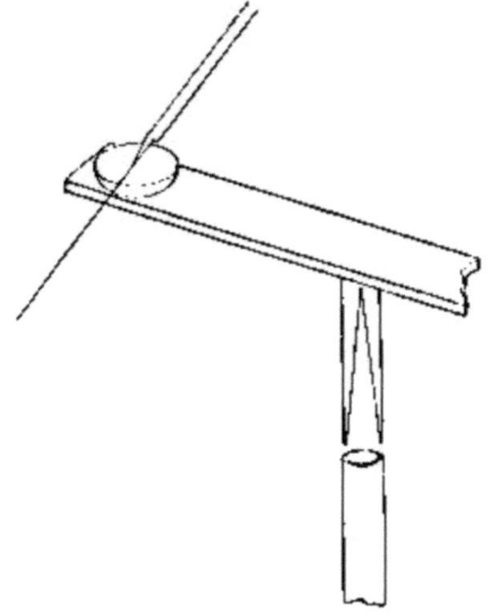

Fig. 74

The washing part of the process is, in the opinion of the writer, the most difficult part of the whole business, and it requires to be very thorough, or the thread will end by drawing out of the solder. In many cases it is better to try to do without any application of nitric acid at all, but, of course, this involves silvering and coppering to exact distances from the ends of the thread — at all events, in apparatus where the effective length of the thread is narrowly prescribed.

It is important not to leave the active parts of the thread appreciably silvered, for the sake of avoiding zero changes due to the imperfect elasticity of the silver. In this soldering process ordinary tinman's solder may be employed; it must be applied very free from dust or oxide.

91. Other Modes of soldering Quartz.

Thick rods of quartz may be treated for attachment by solder in the same way as glass was treated by Professor Kundt to get a foundation for his electrolytically deposited prisms. *[Footnote: See Appendix at end of book.]*

The application of a drop of a strong solution of platinum tetrachloride to the rod will, on drying, give rise to a film of the dry salt, and this may be reduced in the luminous gas flame. During the process, however, the quartz is apt to get rotten, especially if the temperature has been anything approaching a full red heat. The resulting platinum deposit adheres very strongly to the quartz, and may be soldered to as before. This method has been employed by the writer with success since 1887, and may even be extended to thick threads.

It was also found that fusible metal either stuck to or contracted upon clean quartz so as to make a firm joint. In the light of M. Margot's researches (already described), it occurred to me that perhaps my experience was only a special case of the phenomena of adhesion investigated with so much success by M. Margot. I therefore tried whether the alloy of tin and zinc used for soldering aluminium would stick to quartz, and instantly found that this was indeed the case.

Adhesion between the alloy and perfectly clean quartz takes place almost without rubbing. A rod of quartz thus "tinned" can be soldered up to anything to which solder will stick, at once. On applying the method to thick quartz threads, success was instantaneous (the threads were some preserved for ordinary galvanometer suspensions); but when the method was applied to very fine threads, great difficulty in tinning the threads was experienced. The operation is best performed by having the alloy on the end of an aluminium soldering bit, and taking care that it is *perfectly free from oxide* before the thread is drawn across it. There was no difficulty in soldering a thread "tinned" in this manner to a copper wire with tinman's solder, and the joint appeared perfect, the thread breaking finally at about an inch away from the joint.

I allow Mr. Boys' method to stand as I have written it, simply because I have not had time as yet to make thorough tests of the durability of "Margot" joints on the finest threads; but I have practically no doubt as to its perfect applicability, provided always that the solder can be got clean enough when melted on the bit. Very fine threads will require to be stretched before tinning, in order to enable them to break through the capillary barrier of the surface of the melted solder.

92. Soldering. —

It is almost unfair to the arts of the glass-blower or optician to describe them side by side with the humble trade of soldering. Nevertheless, no accomplishment of a mechanical kind is so serviceable to the physicist as handiness with the soldering bit; and, as a rule, there is no other exercise in which the average student shows such lamentable incapacity. The following remarks on the subject are therefore addressed to persons presumably quite ignorant of the way in which soldering is carried out, and do not profess to be more than of the most elementary character.

For laboratory purposes three kinds of solder are in general sufficient. One is the ordinary tinman's solder composed of lead and tin. The second is "spelter," or soft fusible brass, and the third is an alloy of silver and brass called silver solder.

Tinman's solder is used for most purposes where high temperatures are not required, or where the apparatus is intended to be temporary. The "spelter," which is really only finely granulated fusible brass, is used for brazing iron joints. The silver solder is convenient for most purposes where permanency is required, and is especially suited to the joining of small objects.

93. Soft tinman's solder is made by melting together two parts of grain tin and one of soft lead — the exact proportions are not of consequence — but, on the other hand, the purer the constituents the better the solder. Within certain limits, the greater the proportion of tin the cleaner and more fusible is the solder. It is usually

worth while to prepare the solder in the laboratory, for in this way a uniform and dependable product is assured. Good soft lead is melted in an iron ladle and skimmed; the temperature is allowed to rise very little above the melting-point. The tin is then added little by little, the alloy stirred vigorously and skimmed, and sticks of solder conveniently cast by sweeping the ladle over a clean iron plate, so as to pour out a thin stream of solder. If the solder be properly made it will have a mat and bright mottled surface, and will "crackle" when held up to the ear and bent.

Perhaps the chief precaution necessary in making solder is to exclude zinc. The presence of a very small percentage of this metal entirely spoils the solder for tinman's work by preventing its "running" or flowing smoothly under the soldering bit.

Fig.

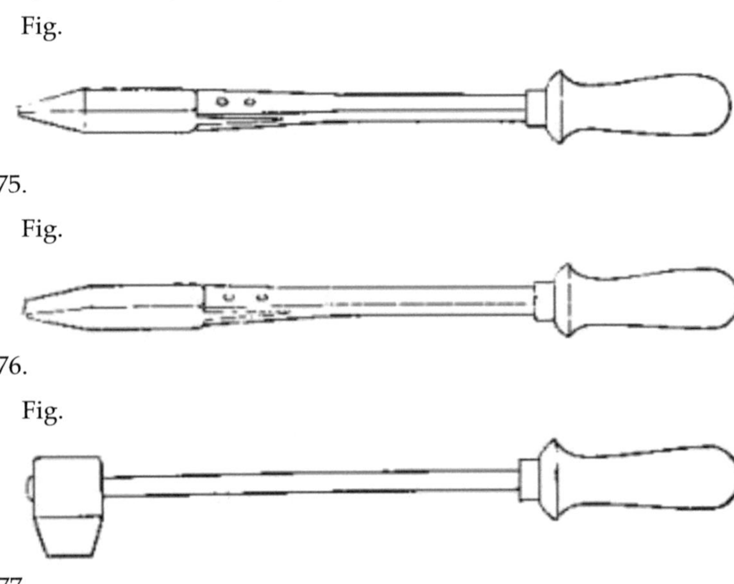

75.

Fig.

76.

Fig.

77.

94. Preparing a Soldering Bit. —

The wedge-shaped edge of one of the forms of bit shown in the sketch is filed to shape and the bit heated in a fire or on a gas heater. A bit of rough sandstone, or even a clean soft brick, or a bit of tin plate having some sand sprinkled over it, is placed in a convenient position and sprinkled with resin.

As soon as the bit is hot enough to melt solder it is withdrawn and a few drops of solder melted on to the brick or its equivalent. The iron or bit is then rubbed to and fro over the solder and resin till the former adheres to and tins the copper head. It will be found advisable to tin every side of the point of the bit and to carry the tinning back at least half an inch from the edge.

If the solder obstinately refuses to adhere, the cause is to be sought in the oxidation of the copper, or of the solder, or both — in either case the result of too high a temperature or too prolonged heating. The simple remedy is to get the iron hot, and then to dress it with an old file, so as to expose a bright surface, which is instantly passed over the resin as a means of preserving it from oxidation. If the process above described be now carried out, it will be found that the difficulty disappears.

Before using the iron, wipe off any soot or coke or burned resin by means of an old rag. An iron tinned in this way is much to be preferred to one tinned by means of chloride of zinc.

A shorter and more usual method is carried out as follows: The solution of chloride of zinc is prepared by adding bits of zinc to some commercial hydrochloric acid diluted with a little (say 25 per cent) of water. The acid may conveniently be placed in a small glazed white jar (a jam pot does excellently), and this should only be filled to about one-quarter of its capacity. An excess of zinc may be added.

It may be fancy, but I prefer a soldering solution made in this way to a solution of chloride of zinc bought as a chemical product. The jar is generally mounted on a heavy leaden base, so as to avoid any danger of its getting knocked over, for nothing is so nasty or bad for tools as a bench on which this noxious liquid has been upset (Fig. 78).

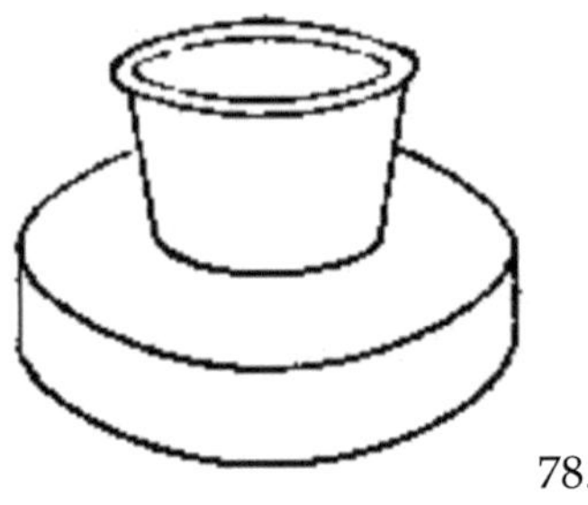

Fig. 78.

To tin a soldering bit, a little of the fluid is dipped out of the jar on to a bit of tin plate bent up at the edges--a few drops is sufficient — and the iron is heated and rubbed about in the liquid with a drop of solder. If the iron is anything like clean it will tin at once and exhibit a very bright surface, but quite dirty copper may be tinned by dipping it for a moment in the liquid in the pot and then working it about over the solder. An iron so tinned remains covered with chloride of zinc, and this must be carefully wiped off if it is intended to use the iron with a resin or tallow flux in lead soldering.

One disadvantage of this process is that the copper bit soon gets eaten into holes and requires to be dressed up afresh. On the other hand, an iron so tinned always presents a nice clean solder surface until the next time it is heated, when it generally becomes very dirty and requires to be carefully wiped before using.

In my experience also an iron so tinned is more easily spoiled as to the state of its surface, "detinned," in fact, by overheating than when the tinning is carried out by resin and friction. When this happens, the shortest way out of the difficulty is the application of the old file so as to obtain a perfectly fresh surface. No one who knows his business ever uses an iron that is not perfectly clean and well tinned.

The iron may be cleaned from time to time by heating it red hot and quenching it in water to get rid of the oxide, which scales off in the process.

95. Soft Soldering. —

In the laboratory the chief application of the process is to copper soldering during the construction of electrical apparatus and to zinc soldering for general purposes.

In ninety-nine cases out of every hundred where difficulties occur their origin is to be traced to dirt. There seems to be some inexplicable kink in the human mind which renders it callous to repeated proofs of the necessity for cleaning surfaces which it is intended to solder. The slightest trace of albuminous or gelatinous matter or shellac will prevent solder adhering to most metals and the same remark applies in a measure to the presence of oxides, although these may be removed by chloride of zinc or prevented from forming by resin or tallow. A touch with an ordinarily dirty hand — I refer to a solderer's hand — will often soil work sufficiently to make the adherence of solder difficult.

The fluxes most generally employed are tallow for lead, resin or Venice turpentine for copper, chloride of zinc for anything except lead, which never requires it. The latter flux has the property (also possessed by borax at a red heat) of dissolving any traces of oxide which may be formed, as well as acting as a protecting layer to the metal.

We may now turn to the consideration of a simple case of soldering, say the joining of two copper wires. The wires are first cleaned either by dipping in a bath of sulphuric and nitric acids — a thing no laboratory should be without — or by any suitable mechanical means. The cleaned wires are then twisted together — there is a regulation way of doing this, but it presents no advantage in laboratory practice — and the joint is sprinkled over with resin, or painted with a solution of resin in alcohol.

The iron, being heated and floated with solder, is held against the joint, the latter being supported on a brick, and the solder is allowed to "sweat" into the joint. Enough solder must be present to penetrate right through the joint. Nothing is gained by rubbing violently with the iron. If the copper is clean it will tin, and if it is dirty it won't, and there the matter ends.

Beginners generally use too small or too cold a bit, and produce a ragged, dirty joint in consequence. If the saving of time be an object, the joint may be twisted together on ordinarily dirty oxidised wires and heated to, say, 200 C. It is then painted with chloride of zinc and soldered with the bit.

There is a difference of opinion as to the relative merits of chloride of zinc and of resin as a flux in soldering copper. Thus the standing German practice is, or was, to employ the former flux in every case for soldering electric light wires, while in England the custom used to be to specify that soldering should be done by resin, and this custom may still prevail; it lingers in Australia at all events.

However, it is agreed on all hands that when chloride of zinc is used it must be carefully washed off. I have known of an electrical engineer insisting on his workmen "licking" joints with their tongues to ensure the total removal of chloride of zinc; it has a horrible taste; and I have occasionally pursued the same plan myself when the soldering of fine wires was in question.

In any case, it is very certain that chloride of zinc left in a joint will ruin it sooner or later by loosening the contact between copper and solder.

Very often it is requisite to solder together two extensive flat surfaces — for instance, in "chucking" certain kinds of brass work. The surfaces to be soldered must be carefully tinned, most conveniently by the help of the blow-pipe and chloride of zinc. After tinning, the surfaces are laid together and heated so as to "sweat" them together; the phrase, though inelegant, is expressive.

96. Soldering Tin Plate. —

If the plate be new and clean, a little resin or its solution in alcohol is all that is necessary as a flux. If the tin plate is rusty the rust must be removed and the clean iron, or rather mild steel, surface exposed. The use of chloride of zinc is practically essential in this case. Tin plate is often spotted with rust long before it becomes rusty as a whole, when, of course, it may be regarded as worn out, and such rust spots are most conveniently removed by means of the plumber's shave-hook. The shave-hook is merely a peculiarly shaped hard steel scraping knife on a handle (Fig. 79).

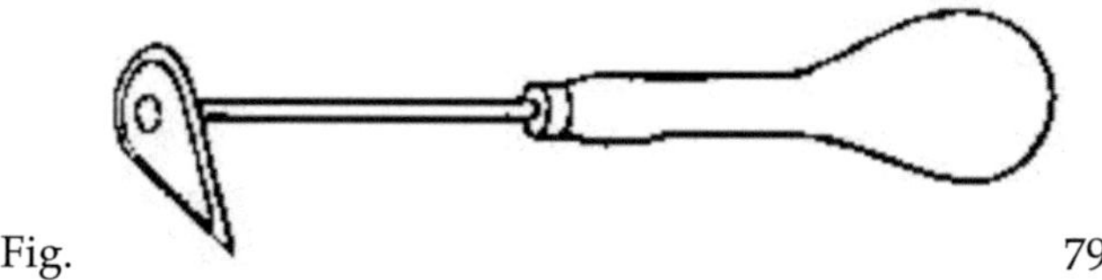

Fig.   79

With tin plate the soldering of long joints is often necessary. The plate must be temporarily held in position either by binding with iron wire, fastening by clamps, or holding by an assistant. The flux is applied and the iron run slowly along the joint. Enough solder is used to completely float the tip of the iron. By arranging the joint so that it slopes downward slightly, and commencing at the upper end, the solder may be caused to flow after the iron, and will leave a joint with the minimum permissible amount of solder in it. By regulating the slope, heat of iron, etc., any desired quantity of solder may be run into the joint.

97. Soldering Zinc. —

Zinc alloys with soft solder very easily, and by so doing entirely spoils it, making, it "crumbly," dirty, and preventing it running. Consequently, in soldering up zinc great care must be taken to prevent the solder becoming appreciably contaminated by the zinc. To this end the zinc surfaces are cleaned by means of a little hydrochloric acid, which is painted on instead of chloride of zinc. Plenty of solder is melted on to the work, and is drawn along over the joint by a single slow motion of the soldering bit. The iron must be just hot enough to make the solder flow freely, and it must never be rubbed violently on the zinc or allowed to linger in one spot; the result of the latter action will be to melt a hole through the zinc, owing to the tendency of this metal to form an easily fusible alloy with the solder.

The art of soldering zinc is a very useful one in the laboratory. The majority of physicists appear to overlook the advantages of zinc considered as a material for apparatus construction. It is light, fairly strong, cheap, easily fusible, and yet hard and elastic when cold. It may be worked as easily as lead at a temperature of, say, 150 to 200 C., and slightly below the melting-point (423 C.) it is brittle and may & powdered. The property of softening at a moderate temperature is invaluable as a means of flattening zinc plate or shaping it in any way. During the work it may be held by means of an old cloth. Zinc sheet which has been heated between iron plates and flattened by pressure retains its flatness very fairly well after cooling.

98. Soldering other Metals —

Iron. —

The iron must be filed clean and then brushed with chloride of zinc solution. Some people add a little sat ammoniac to the chloride of zinc, but the improvement thus made is practically inappreciable. If the iron is clean it tins quite easily, and the process of soldering it is perfectly easy and requires no special comment.

Brass. —

The same method as described for iron succeeds perfectly. The brass, if not exceedingly dirty, may be cleaned by heating to the temperature at which solder melts (below 200 C.), and painting it over with chloride of zinc, or dipping it in the liquor. If now the brass be heated again in the blow-pipe flame, it will be found to tin perfectly well when rubbed over with solder.

German Silver, Platinoid, Silver, and Platinum

are treated like iron. With regard to silver and platinum the same precautions as recommended in the case of zinc must be observed, for both these metals form fusible alloys with solder.

Gold

when pure requires no flux. Standard gold, which contains copper, solders better with a little chloride of zinc.

Lead

must be pared absolutely clean and then soldered quickly with a hot iron, using tallow as a flux. Since solder if over hot will adhere to lead almost anywhere, plumbers are in the habit of specially soiling those parts to which it is not intended that solder shall adhere. The "soiling" paint consists of very thin glue, called size, mixed with lampblack; on an emergency a raw potato may be cut in half, and the work to be soiled may be rubbed over with the cut surface of the potato.

may be soldered after coating with copper by an electrolytic process, as will be described.

99. *Brazing.*

Soldering at a red heat by means of spelter is called brazing. Spelter is soft brass, and is generally made from zinc one part, copper one part; an alloy easily granulated at a red heat; it is purchased in the granular form.

The art of brazing is applied to metals which will withstand a red heat, and the joints so soldered have the strength of brass.

The pieces to be jointed by this method must be carefully cleaned and held in their proper relative positions by means of iron wire. It is generally necessary to soften iron wire as purchased by heating it red hot and allowing it to cool in the air; if this is not done the wire is usually too hard to be employed satisfactorily for binding.

Very thin wire--i.e. above No. 20 on the Birmingham wire gauge — does not do, for it gets burned through, and perhaps allows the work to fall apart at a critical moment.

The work being securely fastened, the next step is to cover the cleaned parts with flux in order to prevent oxidation. For this purpose "glass borax" is employed. "Glass" borax is simply ordinary borax which has been fused for the purpose of getting rid of water of crystallisation. The glass borax is reduced to powder in an iron mortar, for it is very hard, and is then made up into a cream with a little water. This cream is painted on to the parts of the work which are destined to receive the solder.

The next step is to prepare the spelter, and this is easily done by mixing it with the cream, taking care to stir thoroughly with a flattened iron wire till each particle of spelter is perfectly covered with the borax. The mixture should not be too wet to behave as a granular mass, and may then be lifted on to the work by means of the iron spatula.

Care must be taken to place the spelter on those parts only which are intended to receive it, and when this is done, the joint may be lightly powdered over with the dry borax, and will then be ready for heating.

If the object is of considerable size it is most conveniently heated on the forge; if small the blowpipe is more convenient. In the latter case, place the work on a firebrick, and arrange two other bricks on edge about it, so that it lies more or less in a corner. A few bits of coke may also be placed on and about the work to increase the temperature by their combustion, and to concentrate the flame and prevent radiation. The temperature is gradually raised to a bright red heat, when the spelter will be observed to fuse or "run," as it is technically said to do.

If the cleaning and distribution of flux has been successful, the spelter will "run" along the joint very freely, and the work should be tapped gently to make sure that the spelter has really run into the joint. The heating may be interrupted when the spelter is observed to have melted into a continuous mass. As soon as the work has fallen below a red heat it may be plunged into water, a process which has the effect of cracking off the glass-like layer of borax.

There is, however, some risk of causing the work to buckle by this violent treatment, which must of course be modified so as to suit the circumstances of the case. If the joint is in such a position that the borax cannot be filed off, a very convenient instrument for its removal by scraping is the watchmaker's graver, a square rod of hard steel ground to a bevelled point (Fig. 80).

Fig. 80.

Several precautions require to be mentioned. In the first place, spelter is merely rather soft brass, and consequently it often cannot be fused without endangering the rest of the work. A good protection is a layer of fireclay laid upon the more delicate parts, such for instance as any screwed part.

Gun-metal and tap-metal do not lend themselves to brazing so readily as iron or yellow brass, and are usually more conveniently treated by means of silver solder.

Spelter tends to run very freely when it melts, and if the brass surface in the neighbourhood of the joint is at all clean, may run where it is not wanted. Of course some control may be exercised by "soiling" with fireclay or using an oxidising flame; but the erratic behaviour of spelter in this respect is the greatest drawback to its use in apparatus construction. The secret of success in brazing lies in properly cleaning up the work to begin with, and in disposing the borax so as to prevent subsequent oxidation.

100. Silver Soldering. —

This process resembles that last described, but instead of spelter an alloy of silver, copper, and zinc is employed. The solder, as prepared by jewellers to meet special cases, varies a good deal in composition, but for the laboratory the usual proportions are —

For soft silver solder

> Fine silver 2 parts

> Brass wire 1 part

For hard silver solder

> Sterling silver 3 parts

> Brass wire 1 part

The latter is, perhaps, generally the more convenient.

Silver solders may, of course, be purchased at watchmakers' supply shops, and as thus obtained, are generally in thin sheet. This is snipped fine with a pair of shears preparatory to use.

As odds and ends of silver (from old anodes and silver residues) generally accumulate in the laboratory, it is often more convenient to make the solder one's self. In this case it must be remembered in making hard solder by the second receipt that standard silver contains about one-twelfth of its weight of copper — exactly 18 parts copper to 220 silver.

The silver is first melted in a plumbago crucible in a small furnace together with a little borax; if any copper is required this is then added, and finally the brass is introduced. When fusion is complete, the contents of the crucible are poured into any suitable mould.

The quickest and most convenient way of preparing the alloy for use is to convert it into filings with the assistance of a coarse file, or by milling it, if a milling machine is available.

Equal volumes of filings and powdered glass borax are made into a thin paste with water, and applied in an exactly similar manner to that described under the head of "brazing." In fact all the processes there described may be applied equally to the case under discussion, the substitution of silver for spelter being the only variation.

The silver solder is more manageable than spelter, and does not tend to run wild over the work: a property which makes it much more convenient both for delicate joints and in cases where it is desired to restrict the solder to a single point or line. Small objects are almost invariably soldered with silver solder, and are held by forceps or on charcoal in the pointed flame of an ordinary blow-pipe.

101. On the Construction of Electrical Apparatus - Insulators. —

It is not intended to deal in any way with the design of special examples of electrical apparatus, but merely to describe a rather miscellaneous set of materials and processes constantly required in its construction.

It is not known whether there is such a thing as a perfect insulator, even if we presuppose ideal circumstances. Materials as they exist must be regarded merely as of high specific resistance, that is if we allow ourselves to use such a term in connection with substances, conduction through which is neither independent of electromotive force per unit length, nor of previous history.

Even the best of these substances generally get coated with a layer of moisture when exposed to the air, and this as a rule conducts fairly well. Very pure crystalline sulphur and fused quartz suffer from this defect less than any other substances with which the writer is acquainted, but even with them the surface conductivity soon grows to such an extent as totally to mask the internal conduction.

It is proposed to give a brief account of the properties of some insulating substances and their application in electrical construction, and at the same time to indicate the appliances and methods requisite for working them.

With regard to the specific resistances which will be quoted, the numbers must not be taken to mean too much, partly for the reason already given. It is also in general doubtful whether sufficient care has been taken to distinguish the body from the surface conductivity, and consequently numerical estimates are to be regarded with suspicion. The question of "sampling" also arises, for it must be remembered that a change in composition amounting to, say, 1/10000 per cent may be accompanied by a million-fold change in specific resistance.

102. Sulphur. —

This element exists in several allotropic forms, which have very different electric properties. After melting at about 125 C., and annealing at 110 for several hours, the soluble crystalline modification is formed. After keeping for some days — especially if exposed to light — the crystals lose their optical properties, but remain of the same melting-point, and are perfectly soluble in carbon bisulphide. The change is accompanied by a change in colour, or rather in brightness, as the transparency changes.

The "specific resistance" of sulphur in this condition is above 10^{28} C.G.S.E.M. units, or 10^{13} megohms per cubic centimetre for an electric intensity of say 12,000 volts per centimetre. This is at ordinary temperatures. At 75 C. the specific resistance falls to about 10^{25} under similar conditions as to voltage.

In all cases the conductivity appears to increase with the electric intensity, or at all events with an increase in voltage, the thickness of the layer of sulphur remaining the same.

The specific inductive capacity is 3.162 at ordinary temperatures, and increases very slightly with rise of temperature. [Footnote: March 1897. — It is now the opinion of the writer that though the specific inductive capacity of a given sample of a solid element is perfectly definite, yet it is very difficult to obtain two samples having exactly the same value for this constant, even in the case of a material so well defined as sulphur.]

The total residual charge, after ten minutes' charging with an intensity of 12,000 volts per centimetre, is not more than 4 parts in 10,000 of the original charge. In making this measurement the discharge occupied a fraction of a second. The electric strength for a homogeneous plate of crystalline sulphur is not less than 33,000 volts per centimetre, and probably a good deal more. If the sulphur is contaminated with up to 3 per cent of the amorphous variety, as is the case if it is cooled fairly quickly from a temperature of 170 C. or over, the specific resistance falls to from 10^{25} to 10^{26} at ordinary temperatures; and the specific inductive capacity increases up to 3.75, according to the amount of insoluble sulphur present.

The residual charge under circumstances similar to those described above, but with an intensity of about 4000 volts per centimetre is, say, 2 per cent of the initial charge. So far as the writer is aware sulphur is the only solid non-conductor which can be easily obtained in a condition of approximate purity and in samples sufficiently exactly comparable with one another; it is the only one, therefore, that repays any detail of description.

Very pure sulphur can be bought by the ton if necessary from the United Alkali Company of Newcastle-on-Tyne. It is recovered from sulphur waste by the Chance process, which consists in converting the sulphur into hydrogen sulphide, and burning the latter with insufficient air for complete combustion. The sulphur is thrown out of combination, and forms a crystalline mass on the walls and floor of the chamber.

The sulphur which comes into the market consists of this mass broken up into convenient fragments. In order to purify it sufficiently for use as an insulator, the sulphur may be melted at a temperature of 120 to 140 C., and filtered through a plug of glass wool in a zinc funnel; as thus prepared it is an excellent insulator. To obtain the results mentioned in the table it is, however, necessary to conduct a further purification (chiefly from water) by distillation in a glass retort.

The sulphur thus obtained may be cast of any desired form in zinc moulds, the castings and moulds being immediately removed to an annealing oven at a temperature of from 100 to 110 C., where they are left for several hours. If the sulphur is kept melted for some time at 125 C. the annealing is not so important.

The castings may be removed from the mould by slightly heating the latter, but many breakages result. Insulators made on this plan are much less affected by the condensation of moisture from the air than anything except fused quartz. They are, however, very weak mechanically, and apt to crack by exposure to such changes of temperature as go on from day to day. It is clear, however, that in spite of this their magnificent electrical properties fit them for many important uses.

If the sulphur be cooled rapidly from 170 C. or over, a mixture of the crystalline and amorphous varieties of sulphur is obtained. This mixture is very much stronger and tougher than the purely crystalline substance, and may be worked with ordinary hardwood tools into fairly permanent plates, rods, etc. Sheets of pure thick filter paper may also be dipped into sulphur at 170 C., at which temperature air and moisture are mostly expelled, and such sheets show a very considerable insulating power. The sulphur does not penetrate the paper, which therefore merely forms a nucleus.

Cakes of the crystalline or mixed varieties may be made by grinding up some purified sulphur, moistening it with redistilled carbon bisulphide, or toluene, or even benzene (C_6H_6), and pressing it in a suitable mould under the hydraulic press. The plates thus formed are porous, but are splendid insulators, especially if made from the crystalline variety of sulphur, and they appear to keep their shape very well, and do not crack with ordinary temperature changes.

The metals which resist the action of sulphur best are gold and aluminium; while platinum and zinc are practically unacted upon at temperatures below a red heat — in the former case, — and below the boiling-point of sulphur in the latter.

A very convenient test of the purity of sulphur is the colour assumed by it when suddenly cooled from the temperature at which it is viscous. Quite pure sulphur remains of a pale lemon yellow under this treatment, but the slightest trace of impurity, such as arises from dust containing organic matter, stains the sulphur, and renders it darker in colour.

103. *Fused Quartz. —*

This is on the whole the most reliable and most perfect insulator for general purposes. No exact numerical data have been obtained, but the resistivity must certainly be of the same order as that of pure sulphur at its best. The influence of the moisture of the air also reaches its minimum in the case of quartz, as was originally observed by Boys.

As yet, however, the material can only be obtained in the form of rods or threads. For most purposes rods of about one-eighth of an inch in diameter are the most convenient. These rods may be used as insulating supports, and succeed perfectly even if they interpose less than an inch of their length to electrical conduction. The sketch (Figs. 81 and 81A) shows (to a scale of about one-quarter full size) a complete outfit for elementary electrostatic experiments, such as has been in use in the writer's laboratory for five years. With these appliances all the fundamental experiments may be performed, and the apparatus is always ready at a moment's notice.

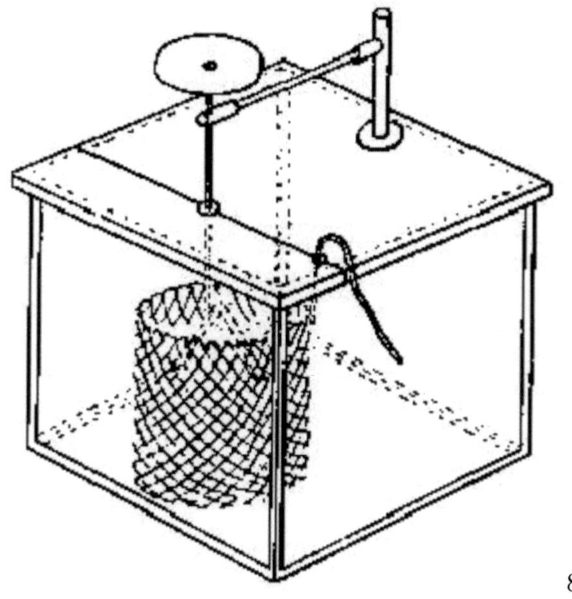

Fig. 81.

Though quartz does not condense moisture or gas to form a conducting layer of anything like the same conductivity as in the case of glass or ebonite, still it is well to heat it if the best results are to be obtained. For this purpose a small pointed blow-pipe flame may be used, and the rods may be got red-hot without the slightest danger of breaking them. They then remain perfectly good and satisfactory for several hours at least, even when exposed to damp and dusty air.

The rods are conveniently held in position by small brass ferrules, into which they are fastened by a little plaster of Paris. Sealing-wax must be avoided, on account of the inconvenience it causes when the heating of the rods is being carried out.

One useful application of fused quartz is to the insulation of galvanometer coils (Fig. 82), another to the manufacture of highly insulating keys (Fig. 83); while as an insulating suspension it has all the virtues. If it is desired to render the threads conducting they may be lightly silvered, and will be found to conduct well enough for electrometer work before the silver coating is thick enough to sensibly impair their elastic properties.

Fig.

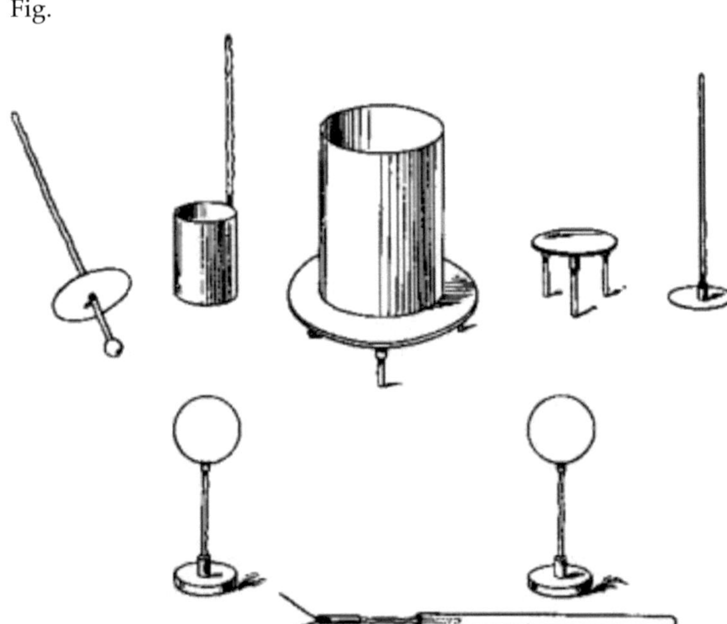

81A.

Fig. 82 is a full-size working drawing of a particular form of mounting for galvanometer coils. The objects sought to be attained are

(1) high insulation of the coils,

(2) easy adjustment of the coils to the suspended system.

The first object is attained as follows. The ebonite ring A is bored with four radial holes, through which are slipped from the inside the fused quartz bolt-headed pins B. The coil already soaked in hard paraffin is placed concentrically in the ring A by means of a special temporary centering stand. The space between the coil and the ring is filled up with hard paraffin, and this holds the quartz pins in position. The system of ebonite ring, coil, and pins is then fastened into the gun-metal coil carrier, which is cut away entirely, except near the edges, where it carries the pin brackets C. These brackets can swivel about the lower fastening at E before the latter is tightened up.

The coil is now adjusted in the adjusting stand to be concentric with the axis of symmetry of the coil carrier, and the supporting pins are slipped into slot holes cut in the brackets, the brackets being swivelled as much as necessary to allow of this. When the pins are all inserted the brackets are screwed up by the screws at E. The pins are then cemented firmly to the brackets by a little plaster of Paris. The coil carrier can now be adjusted to the galvanometer frame by means of screws at D, which pass through wide holes in the carrier and bold the latter in position by their heads. In the sectional plan the parts of the galvanometer frame are shown shaded. The front of the frame at F F is of glass, and the back of the frame is also made of glass, though this is not shown in the section.

A represents an ebonite ring into which the wire coil is cemented by means of paraffin. B B B B are quartz pins, with heads inside the ebonite ring. C C C are slotted brackets adjustable to the pins and capable of rotation by releasing the screws E E. D D are the screws holding the coil carriage to the galvanometer framework. These screws pass through large holes in the carriage so as to allow of some adjustment.

Fig.

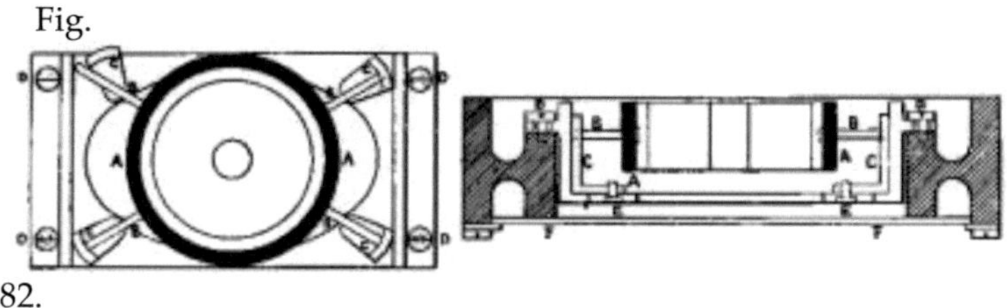

82.

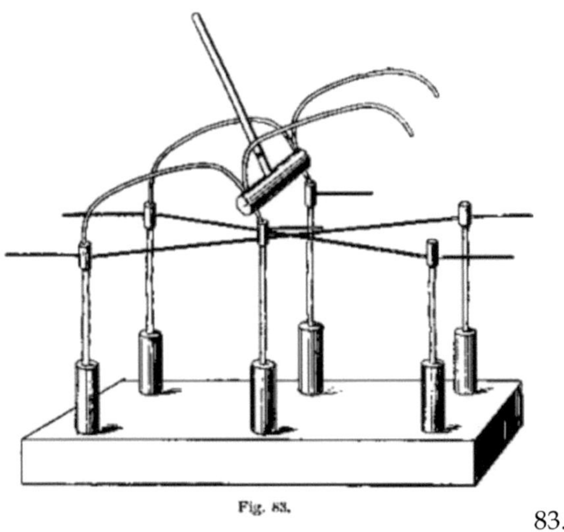

Fig. 83.

104. *Glass.* —

When glass is properly chosen and perfectly dry it has insulating properties possibly equal to those possessed by quartz or crystalline sulphur. For many purposes, however, its usefulness is seriously reduced by the persistence with which it exhibits the phenomena of residual charge, and the difficulty that is experienced in keeping it dry.

The insulating power of white flint glass is much in excess of that of soft soda glass, which is a poor insulator, and of ordinary green bottle glass. The jars of Lord Kelvin's electrometers, which insulate very well, are made of white flint glass manufactured in Glasgow, but it is found that occasionally a particular jar has to be rejected on account of its refusing to insulate, and this, if I understand aright, even when it exhibits no visible defects.

A large number of varieties of glass were tested by Dr. Hopkinson at Messrs. Chance Bros. Works, in 1875 and 1876 (Phil. Trans., 1877), and in 1887 (Proc. Roy. Soc. xli. 453), chiefly with a view to the elucidation of the laws regulating the residual charge; and incidentally some extraordinarily high insulations were noted among the flint glasses. The glass which gave the smallest residual charge was an "opal" glass; and flint glasses were found to insulate 10^5

times as well as soda lime glasses. The plates of Wimshurst machines are made of ordinary sheet window glass, but as the insulating property of this material appears to vary, it is generally necessary to clean, dry, and test a sheet before using it. With regard to hard Bohemian glass, this is stated by Koeller (Wien Bericht) to insulate ten times as well as the ordinary Thuringian soft soda glass.

On the whole the most satisfactory laboratory practice is to employ good white flint glass. The only point requiring attention is the preparation of the glass by cleaning and drying. Of course all grease or visible dirt must be removed as described in an earlier chapter (13), but this is only a beginning. The glass after being treated as described and got into such a state as to its surface that clean water no longer tends to dry off unequally, must be subjected to a further scrub with bibulous paper and a clear solution of oleate of soda. The glass is then to be well rinsed with distilled water and allowed to drain on a sheet of filter paper.

A very common cause of failure lies in the contamination of the glass with grease from the operator's fingers. Before setting out to clean the glass the student will do well to wash his hands with soap and water, then with dilute ammonia and finally with distilled water.

In the case of an electrometer jar which has become conducting but is not perceptibly dirty, rubbing with a little oleate of soda and a silk ribbon, followed, of course, by copious washing, does very well. If there is any tin-foil on the jar, great care must be taken not to allow the glass surface to become contaminated by the shellac varnish or gum used to stick the tin-foil in position.

Finally, the glass should be dried by radiant heat and raised to a temperature of 100 C. at least, and kept at it for at least half an hour. Before drying it is of course advisable to allow the water to drain away as far as possible, and if the water is only the ordinary distilled water of the laboratory, the glass is preferably wiped with a clean bit of filter paper; any hairs which may be left upon the glass will brush off easily when the glass is dry.

In order to obtain satisfactory results the glass must be placed in dry air before it has appreciably cooled. This is easily done in the case of electrometer jars, and so long as the air remains perfectly dry through the action of sulphuric acid or phosphorus pentoxide, the jar will insulate. The slightest whiff of ordinarily damp air will, however, enormously reduce the insulating power of the glass, so that unvarnished glass surfaces must be kept for apparatus which is practically air-tight.

For outside or imperfectly protected uses the glass does better when varnished. It is a fact, however, that varnished glass is rarely if ever so good as unvarnished glass at its best. Too much care cannot be taken over the preparation of the varnish; French polish, or carelessly made shellac varnish, is likely to do more harm than good.

The best orange shellac must be dissolved in good cold alcohol by shaking the materials together in a bottle. The alcohol is made sufficiently pure by starting with rectified spirit and digesting it in a tin flask over quick-lime for several days, a reversed condenser being attached. A large excess of lime must be employed, and this leads to a considerable loss of alcohol, a misfortune which must be put up with.

After, say, thirty hours' digestion, the alcohol may be distilled off and employed to act on the shellac. In making varnish, time and trouble are saved by making a good deal at one operation — a Winchester full is a reasonable quantity. The bottle may be filled three-quarters full of the shellac flakes and then filled up with alcohol; this gives a solution of a convenient strength.

The solution, however, is by no means perfect, for the shellac contains insoluble matter, and this must be got rid off.`' One way of doing this is to filter the solution through the thick filtering paper made by Schleicher and Schuell for the purpose, but the filtering is a slow process, and hence requires to be conducted by a filter paper held in a clip (not a funnel) under a bell jar to avoid evaporation.

Another and generally more convenient way in the laboratory is to allow the muddy varnish to settle — a process requiring at least a month — and to decant the clear solution off into another bottle, where it is kept for use. The muddy residue works up with the next lot of shellac and alcohol, which may be added at once for future use.

The glass to be varnished is warmed to a temperature of, say, 50 C., and the varnish put on with a lacquering brush; a thin uniform coat is required. The glass is left to dry long enough for the shellac to get nearly hard and to allow most of the alcohol to evaporate. It is then heated before a fire, or even over a Bunsen, till the shellac softens and begins to yield its fragrant characteristic smell.

If the coating is too heavy, or if the heating is commenced before the shellac is sufficiently dry, the latter will draw up into "tears," which are unsightly and difficult to dry properly. On no account must the shellac be allowed to get overheated. If the varnish is not quite hard when cold it may be assumed to be doing more harm than good.

In varnishing glass tubes for insulating purposes it must be remembered that the inside of the tube is seldom closed perfectly as against the external air, and consequently it also requires to be varnished. This is best done by pouring in a little varnish considerably more dilute than that described, and allowing it to drain away as far as possible, after seeing that it has flooded every part of the tube.

During this part of the process the upper end of the tube must be closed, or evaporation will go on so fast that moisture will be deposited from the air upon the varnished surface. Afterwards the tube may be gently warmed and a current of air allowed to pass, so as to prevent alcohol distilling from one part of the tube to another. The tube is finally heated to the softening point of shellac, and if possible closed as far as is practicable at once.

105. Ebonite or Hard Rubber. —

This exceedingly useful substance can be bought of a perfectly useless quality. Much of the ebonite formerly used to cover induction coils for instance, deteriorates so rapidly when exposed to the air that it requires to have its surface renewed every few weeks.

The very best quality of ebonite obtainable should be solely employed in constructing electric works. It is possible to purchase good ebonite from the Silvertown Rubber Company (and probably from other firms), but the price is necessarily high, about four shillings per pound or over.

At ordinary temperatures ebonite is hard and brittle and breaks with a well-marked conchoidal fracture. At the temperature of boiling water the ebonite becomes somewhat softened, so that it is readily bent into any desired shape; on cooling it resumes its original hardness.

This property of softening at the temperature of boiling water is a very valuable one. The ebonite to be bent or flattened is merely boiled for half an hour or so in water, taken out, brought to the required shape as quickly as possible, and left to cool clamped in position.

The sheet ebonite as it comes from the makers is generally far from flat. It is often necessary to flatten a sheet of ebonite, and of course this is the more easily accomplished the smaller the sheet. Consequently a bit of ebonite of about the required size is first cut from the stock sheet by a hack-saw such as is generally used for metals. This piece is then boiled and pressed between two planed iron plates previously warmed to near 100 C.

With pieces of ebonite such as are used for the tops of resistance boxes, measuring, say, 20 X 8 X 11 inches, very little trouble is experienced. The sheets when cold are found to retain the flatness which has been forced upon them perfectly well. It is otherwise with large thin sheets such as are used for Holtz machines. I have succeeded fairly, but only fairly, by pressing them in a "gluing press," consisting of heavy planed iron slabs previously heated to 100 C.

I do not know exactly how best to flatten very thin and large sheets. It is easy to make large tubes out of sheet ebonite by taking advantage of the softening which ebonite undergoes in boiling water. A wooden mandrel is prepared of the proper size and shape. The ebonite is softened and bent round it; this may require two or three operations, for the ebonite gets stiff very quickly after it is taken out of the water. Finally the tube is bound round the mandrel with sufficient force to bring it to the proper shape and boiled in water, mandrel and all. The bath and its contents are allowed to cool together, so that the ebonite cools uniformly.

Tubes made in this way are of course subject to the drawback of having an unwelded seam, but they do well enough to wind wire upon if very great accuracy of form is not required. If very accurate spools are needed the mandrel is better made of iron or slate and the spool is turned up afterwards. The seam may be strapped inside or at the ends by bits of ebonite acting as bridges, and the seam itself may be caulked with melted paraffin or anthracene.

Working Ebonite. —

Ebonite is best worked as if it were brass, with ordinary brass-turning or planing tools. These tools should be as hard as possible, for the edges are apt to suffer severely, and blunt tools leave a very undesirable woolly surface on the ebonite. In turning or shaping ebonite sheets it is as well to begin by taking the skin off one side first, and then reversing the sheet, finishing the second side, and then returning to the first. This is on account of the fact that ebonite sometimes springs a little out of shape when the skin is removed.

Turned work in ebonite, if well done, requires no sand-papering, but may be sufficiently polished by a handful of its own shavings and a little vaseline. The advantage of using a polished ebonite surface is that such a surface deteriorates more slowly under the influence of light and air than a surface left rough from the tool. If very highly polished surfaces are required, the ebonite after tooling is worked with fine pumice dust and water, applied on felt, or where possible by means of a felt buff on the lathe, and finally polished with rouge and water, applied on felt or cloth.

Ebonite works particularly well under a spiral milling cutter, and sufficiently well under an ordinary rounded planing tool, with cutting angle the same as for brass, and hardened to the lightest straw colour.

It is not possible, on the other hand, to use the carpenter's plane with success, for the angle of the tool is too acute and causes the ebonite to chip.

In boring ebonite the drill should be withdrawn from the hole pretty often and well lubricated, for if the borings jam, as they are apt to do, the heat developed is very great and the temper of the drill gets spoiled. Ebonite will spoil a drill by heating as quickly as anything known; on the other hand, it may be drilled very fast if proper precaution is taken.

It is advisable to expose ebonite to the light as little as possible, especially if the surface is unpolished, for under the combined action of light and air the sulphur at the surface of the ebonite rapidly oxidises, and the ebonite becomes covered with a thin but highly conducting layer of sulphurous or sulphuric acid or their compounds. When this happens the ebonite may be improved by scrubbing with hot water, or washing freely with alcohol rubbed on with cotton waste in the case of apparatus that cannot be dismounted.

A complete cure, however, can only be effected by scraping off the outer layer of ebonite so as to expose a fresh surface. For this purpose a bit of sheet glass broken so as to leave a curved edge is very useful, and the ebonite is then scraped like a cricket bat. In designing apparatus for laboratory use it is as well to bear in mind that sooner or later the ebonite parts will require to be taken down and scraped up. Rods or tubes are, of course, most quickly treated on the lathe with rough glass cloth, and may be finished with fine sandpaper, then pumice dust and water, applied on felt. After cleaning the pumice off by means of water and a rag, the final touch may be given by means of vaseline, applied on cloth or on ebonite shavings.

106. *Mica.* —

A great variety of minerals go under this name. Speaking generally, the Russian micas coming into commerce are potash micas, and mica purchased in England may be taken to be potash mica, especially if it is in large sheets.

At ordinary temperatures "mica" of the kind found in commerce is an excellent insulator. Schultze (*Wied. Ann.* vol. xxxvi. p. 655) comes to the conclusion that both at high and at low temperatures mica (of all kinds?) is a better insulator than white "mirror glass," the composition of which is not stated. The experiments of the author referred to were apparently left unfinished, and altogether too much stress must not be laid on the results obtained, one of which was that mica conducts electrolytically to some extent at high temperatures.

Bouty (*Journal de Physique*, 1890 [9], 288) and J. Curie (*Thèse de Doctorat*, Paris, 1888) agree in making the final conductivity of the mica used in Carpentier's condensers exceedingly small — at all events at ordinary temperatures. Bearing in mind that for such substances the term specific resistance has no very definite meaning, M. Bouty considers it is not less than 3.19×10^{28} E.M. units at ordinary temperatures. M. Bouty gives a note or illustration of what such numbers mean — a precaution not superfluous in cases where magnitudes are denoted logarithmically. Referring to the value quoted, viz. 3.19×10^{28}, M. Bouty says, "Ce serait la resistance d'une colonne de mercure de 1^{mmq} de section et de longueur telle que la lumi航 se propageant dans le vide, mettrait plus de 3000 ans A se transmettre d'une extr魴t頷 I'autre de la colonne."

M. Bouty returns to the study of mica (muscovite) in the *Journal de Physique* for 1892, p. 5, and there deals with the specific inductive capacity, which for a very small period of charge he finds has the value 8 — an enormous value for such a good insulator, and one that it would be desirable to verify by some totally distinct method. This remark is enforced by the fact that M. Klemencic finds the number 6 for the same constant. The temperature coefficient of this constant was too small for M. Bouty to determine. The electric intensity was of the order of 100 volts per centimetre, and the experiments seem to indicate that the specific inductive capacity would be only slightly less if referred to a period of charge indefinitely short.

I have found that the residual charge in a mica condenser, made according to Carpentier's method (to be described below), is about 1 per cent of the original charge under the following circumstances.

Voltage 300 volts on a plate 0.2 mm. thick, duration of charge ten minutes, temperature about 20 C. To get this result the mica must be most carefully dried. This and other facts indicate that the so-called residual charge on ordinary condensers is, to a very large extent, due to the creeping of the charge from the armatures over the more or less conducting varnished surfaces of the mica, and its slow return on discharge.

This source of residual charge was carefully guarded against by Rowland and Nichols (*Phil. Mag.* 1881) in their work on quartz, and is referred to by M. Bouty, who adduces some experiments to show that his own results are not vitiated by it. On the other hand, M. Bouty shows that a small rise in temperature enormously affects the state of a mica surface, and that the surface gets changed in such a way as to become very fairly conducting at 300 C. Also anybody can easily try for himself whether exposing a mica condenser plate which has been examined in presence of phosphorus pentoxide to ordinary air for five minutes will not enormously increase the residual charge, as has always been the case in the writer's experience, and if so, it is open to him to suggest some cause other than surface creeping as an explanation.

M. Bouty, using less perfectly dried mica, did not get so good a result as to smallness of residual charge as the one above quoted.

The chief use of mica for laboratory purposes depends on the ease with which it can be split, and also upon the fact that it may be considerably crumpled and bent without breaking. It therefore makes an excellent dielectric in so far as convenience of construction is concerned in the preparation of condensers, and lends itself freely to the construction of insulating washers or separators of any kind. Its success as a fair insulator at moderate temperatures has led to its use in resistance thermometers, where it appears to have given satisfaction up to, at all events, 400 C.

It is worth a note that according to Werner Siemens, who had immense experience (*Wied. Ann.* vol. clix.), soapstone is the only reliable insulator at a red heat, but, no doubt, a good deal depends on the particular specimen investigated.

107. Use of Mica in Condensers. —

If good results are desired it is essential to select the mica very carefully. Pieces appreciably stained, — particularly if the stain is not uniformly distributed, — cracked pieces, and pieces tending to flake off in patches should be rejected. The best samples of mica that have come under the writer's observation are those sheets sold for the purpose of giving to silver photographic prints that hideous glazed surface which some years ago was so popular.

Sheets of mica about 0.1 to 0.2 mm. thick form good serviceable condenser plates, and will certainly stand a pressure of 300 volts, and most likely a good deal more. The general practice in England seems to have been to build up condensers of alternate sheets of varnished or paraffined-mica and tin-foil.

This practice is open to several objections. In the first place, the capacity of a condenser made in this way varies with the pressure binding the plates together. In the second place, the amount of mica and tin-foil required is often excessive in consequence of the imperfect contact of these substances. Again, the inevitable air film between the mica and tin-foil renders condensers so made unsuitable for use with alternating currents, owing to the heating set up through air discharges, and which is generally, though often (if not always) wrongly, attributed to dielectric hysteresis.

These imperfections are to a great extent got over by M. Carpentier's method of construction, which is, however, rather more costly both in material and labour. On the other hand, wonderful capacities are obtained with quite small amounts of mica. M. Bouty mentions a condenser of one microfarad capacity weighing 1500 grms. and contained in a square box measuring 12 centimetres on the side, and about 3 centimetres thick.

The relation between the capacity and surface of doubly-coated plates is in electro-static units —

Capacity = (sp. ind. capacity X area of one surface)/(4pi X thickness)

This may be reduced to electro-magnetic units by dividing by 9×10^{20}, and to microfarads by further multiplying by 10^{15}.

M. Carpentier begins, of course, by having his mica scrupulously clean and well selected. It is then silvered by one of the silvering processes (65) on both sides, for which purpose the sheets may be suspended in a paraffined wood rack, so as to lie horizontally in the silvering solution, a space of about half an inch being allowed between the sheets. The silvering being finished, the sheets are dipped along two parallel edges in 75 per cent nitric acid. With regard to the third and fourth edges of the sheet, the silver is removed on one side only, using a spun glass brush; if we agree to call the two surfaces of the mica A and B respectively, and the two edges in question C and D, then the silver is removed from the A side along edge C, and from the B side along edge D. The silvered part is shown shaded in Fig. 84. By this arrangement the silver and mica plates may be built up together so as to form the same mutual arrangement of contacts as in an ordinary mica tin-foil condenser.

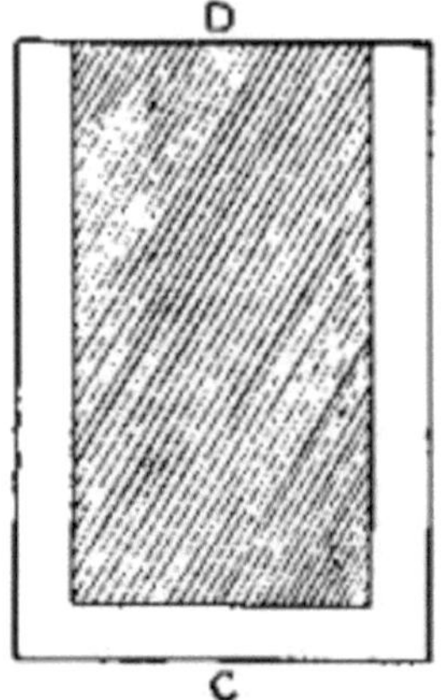
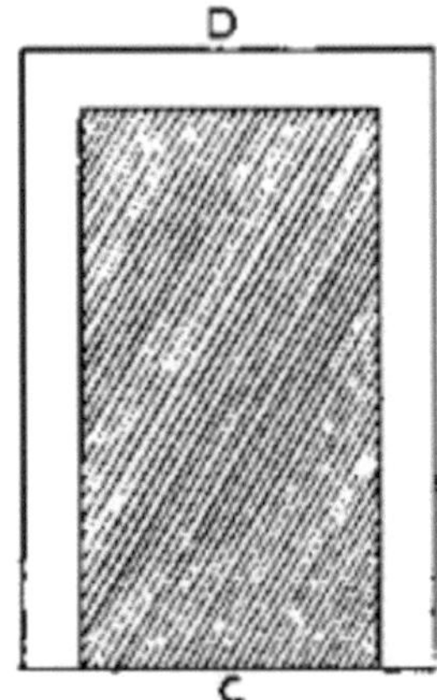

Fig. 84.

It need hardly be said that the sheets require very complete washing after treatment with nitric acid, followed by a varnishing of the edges as already described in the case of glass, and baking at a temperature of 140 C. in air free from flame gases, till the shellac begins to emit its characteristic odour and is absolutely hard when cold.

The plates are then built up so as to connect the sheets which require to be connected, and to insulate the other set. General contact is, if necessary, secured by means of a little silver leaf looped across from plate to plate — a part of the construction which requires particular attention and clean hands, for it is by no means so easy to make an unimpeachable contact as might at first appear.

The condenser, having been built up, may be clamped solid and placed in its case; the capacity will not depend appreciably on the tightness of the clamp screws — a great feature of the construction. Such a condenser will not give its best results unless absolutely dry. I have kept one very conveniently in a vacuum desiccator over phosphorus pentoxide, but if of any size, the condenser deserves a box to itself, and this must be air-tight and provided with a drying reagent, so arranged that it can be removed through a manhole of some sort.

Contact to the brass-work on the lid may be made by pressing spring contacts tightly down upon the ends of the silver foils and carrying the connections through the lid. This also serves to secure the condenser in position.

108. *Micanite.* —

This substance, though probably comparing somewhat unfavourably with the insulators already enumerated, and being subject to the uncertainties of manufacture, has during the last few years achieved a considerable success in American electrical engineering construction. It is composed of scrap mica and shellac varnish worked under pressure to the desired shape, and may be obtained in sheets, plates, and rods, or in any of the forms for which a die happens to have been constructed.

Of course, in special cases it would be worth while to prepare a die, and then the attainable forms would be limited by moulding considerations only. The writer's experience is very limited in this matter, but Dr. Kennelly, with whom he communicated on the subject, was good enough to reply in favour of micanite for engineering work.

109. *Celluloid.* —

This material is composed of nitrocellulose and camphor.

It has fair insulating properties, and may be obtained in a variety of forms, but has now been generally abandoned for electrical work on account of its inflammability.

110. *Paper.*

Pure white filter paper, perfectly dry, is probably a very fair insulator; the misfortune is that in practice it cannot be kept dry. Under the most favourable circumstances its specific resistance may approach 10^{24} E.M. units. It must therefore be considered rather as a partial conductor than as an insulator. The only case of the use of dry paper as an insulator in machine construction which has come under the writer's notice is in building up the commutators of the small motors which used to drive the Edison phonographs.

Its advantages in this connection are to be traced to the fact that a commutator so built up is durable and keeps a clean surface. Of course, the use of paper as an insulator for telephone wires is well known, but its success in this direction depends less upon its insulating properties than upon the fact that it can be arranged in such a way as to allow of the wires being partially air insulated, an arrangement tending to reduce the electrostatic capacity of the wire system.

At one time it was the custom of instrument makers to employ ordinary printed paper in the shape of leaves torn from books or the folios of old ledgers to form the dielectric of the condensers used in connection with the contact breakers of induction coils. This practice has nothing but economy to recommend it, for cases often occur in which the paper, by gradual absorption of moisture from the air, comes to insulate so badly that it practically short circuits the spark gap, and so stops the action of the coil. Three separate cases have come within the writer's experience.

Some measurements of the resistance of paper have been made by F. Uppenborn (*Centralblatt fuer Electrotechnik*, Vol. xi. p. 215, 1889). There is an abstract of the paper also in *Wiedemann's Beiblaetter* (1889, vol. xiii. P. 711). Uppenborn examined the samples of paper under normal conditions as to moisture and obtained the following results: —

Description of Paper	I Pressure Intensity	II. Specific Resistance corresponding to pressures as in Column I. Ohms.	III Pressure Intensity.	IV. Specific Resistance corresponding to Column III. Ohms.
Common cardboard 2.3 mm. thick	0.05 kilo. per 6000 sq. mm.	4.85×10^{15}	20 kg. per 6000 sq. mm.	4.7×10^{14}

Gray paper, 0.26 mm. thick	0.05 kilo. per 5000 sq. mm.	3.1×10^{15}	20 kg. per 5000 sq. mm.	8×10^{14}
Yellow parchment paper-09 mm. thick	0.05 kilo. per 5300 sq. mm.	3.05×10^{16}	20 kg. per 5300 sq. mm.	8×10^{16}
Linen tracing cloth	0.05 kilo. per 6000 sq. mm.	1.35×10^{16}	20 kg. per 33,000 sq. mm.	1.86×10^{15}

111. Paraffined Paper. —

Like wood and other semiconductors, paper can be vastly improved as an insulator by saturating it with melted paraffin. To get the best results a pure paper free from size must be employed — gray Swedish filter paper does well. This is dried at a temperature above 100 C. for, say, half an hour, and the sheets are then floated on the top of paraffin, kept melted at 140 C. or thereabout in a baking dish. As soon as the paper is placed upon the melted paraffin the latter begins to soak through, in virtue of capillary action, and drives before it the air and moisture, causing a strongly marked effervescence.

After about one minute the paper may be thrust below the paraffin to soak. When a sufficient number of papers have accumulated, and when no more gas comes off, the tray may be placed in a vacuum box (Fig. 85), and the pressure reduced by the filter pump. As the removal of the air takes time, provision must be made for keeping the bath hot.

A vacuum may be maintained for about an hour, and air then readmitted. Repeated exhaustions and readmissions of air, which appear to improve wood, do not give anything like such a good result with paper. In using a vacuum box provision must be made in the shape of a cool bottle between the air pump and the box. If this precaution be omitted, and if any paraffin splashes on to the hot surface of the box, it volatilises with decomposition and the products go to stop up the pump. Paraffin with a melting-point of 50 C. or upwards does well.

The bath should be allowed to cool to about 60 C. before the papers are removed, so that enough paraffin may be carried out to thoroughly coat the paper and prevent the entrance of air.

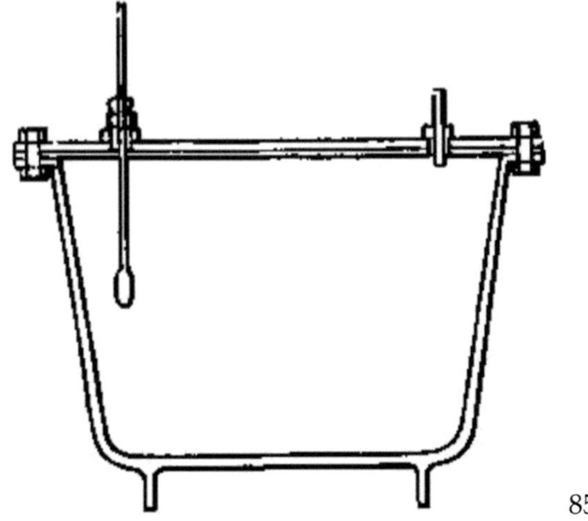

Fig.   85.

Fig. 85 is a section of a vacuum vessel which has been found very convenient. It measures about two feet in diameter at the top. It is round, because it is much easier to turn one circular surface than to plane up four surfaces, which has to be done if the box is square. Both the rim of the vessel and the approximating part of the cover require to be truly turned and smoothly finished. A very good packing is made of solid indiarubber core about half an inch thick. This is carefully spliced — cemented by means of a solution of rubber in naphtha, and the splice sewed by thick thread. The lid ought to have a rim fitting inside the vessel, for this keeps the rubber packing in place; the rim has been accidentally omitted in Fig. 85. The bolts should not be more than five inches apart, and should lie at least half an inch in diameter, and the rim and lid should be each half an inch thick.

Condensers may now be built up of sheets of this prepared paper interleaved with tin-foil in the ordinary way. If good results are required, the condenser when finished is compressed between wooden or glass end-pieces by means of suitable clamps. It can then be put in a box of melted paraffin, heated up to 140 C., and exhausted by means of the water pump for several hours.

In this process the air rushes out from between the paper and foils with such vehemence that the paraffin is generally thrown entirely out of the box. To guard against this the box must be provided with a loosely fitting and temporary lid, pierced with several holes.

The real test as to when exhaustion is complete would be the cessation of any yield of air or water. Since it is not generally convenient to make the vacuum box so air-tight that there are absolutely no leaks at all, and as the paraffin itself is, I think, inclined to "crack" slightly at the temperature of 140 C., this test or criterion cannot be conveniently applied.

Two exhaustions, each of about two hours' duration, have, however, in my experience succeeded very well, provided, of course, that the dielectric is prepared as suggested. At the end of the exhaustion process the clamping screws are tightened as far as possible, the condenser remaining in its bath until the paraffin is pasty.

Condensers made in this way resist the application of alternating currents perfectly, as the following tests will show. The dielectric consisted of about equal parts of hard paraffin and vaseline. A condenser of about 0.123 microfarads capacity and an insulation resistance of 2000 megohms, *[Footnote: As tested by a small voltage.]* having a tin-foil area of 4.23 square metres (about), and double papers each about 0.2 mm. thick, designed to run at 2000 volts with a frequency of 63 complete periods, was tested at this frequency.

The condenser was thoroughly packed all round in cotton-wool to a thickness of 6 inches, and its temperature was indicated more or less by a thermometer plunged through a hole in the lid of the containing box and of the condenser box, and resting on the upper surface of one set of tin-foil electrodes, from which the soft paraffin mixture had been purposely scraped away. The following were the results of a four hours' run at a voltage 50 per cent higher than that for which the condenser was designed — i.e. 3000 volts.

Times.	Vol-	Tempera-	Tempera-	Diffe-

H rs.	Mi n.	tage	ture in Con- denser.	ture in Air.	rence
2	10	2750	22.8 C.	23.0 C.	+ 0.2
3	10	2700	23.0 C.	23.3 C.	+ 0.3
3	18	3200	23.1 C.	23.0 C.	-0.1
4	10	3200	23.3 C.	23.7 C.	+ 0.4
5	10	3100	23.6 C.	23.4 C.	-0.2
6	10	3000	23.8 C.	23.35 C.	-0.45

An idea of the order of the amount of waste may be formed from the following additional experiment.

A condenser similar to the one described was filled with oil of a low insulating power. It was tested calorimetrically, and also by the three voltmeter method, which, however, proved to be too insensitive. The temperature rise in the non-conducting box in air was about 0.3 C. per hour, and the loss of power was found to be less than 0.1 per cent. In the present case the actual rise was only 1 in four hours, and the integral give and take between the condenser and the air is practically nothing; consequently we may consider with safety that the rate of rise is certainly less than 1 degree per three hours. The voltage and frequency were about the same in both experiments, consequently the energy passed is about proportional to the capacity used in the two experiments.

From this it follows that since the specific heat of both condensers was the same (nearly), the loss in the present case is a good deal less than one-tenth per cent. The residual charge is also much less than when the condenser is simply built up of paper paraffined in an unsystematic manner, and from which the air and water have been imperfectly extracted, as by baking the condenser first, and then immersing it in paraffin or oil.

It is usual to consider that the phenomena of residual charge and heating in condensers, to which alternating voltages are applied, are closely allied. This is true, but the alliance is not one between cause and effect — at all events, with regard to the greater part of the heating. The imperfect exclusion of air and moisture, particularly the latter, certainly increases the residual charge by allowing surface creeping to occur; but it also acts directly in producing heating, both by lowering the insulation of the condenser and by allowing of air discharges between the condenser plates.

Of these causes of heating, the discharges in air or water vapour are probably the more important. Long ago a theory of residual charge was given by Maxwell, based on the consideration of a laminated dielectric, the inductivity and resistance of which varied from layer to layer. It was shown that such an arrangement, and hence generally any want of homogeneity in a direction inclined to the lines of force leading to a change of value of the product of specific resistance and specific inductive capacity, would account for residual charge.

This possible explanation has been generally accepted as the actual explanation, and many cases of residual charge attributed to want of homogeneity, which are certainly to be explained in a simpler manner. For instance, the residual charge in a silvered mica plate condenser, carefully dried, can be increased at least tenfold by an exposure of a few minutes to ordinarily damp air. The same result occurs with condensers of paraffined or sulphured paper; and these are the residual changes generally observed. The greater part must be due to creeping.

112. *Paraffin —*

This substance has long enjoyed great popularity in the physical laboratory. Its specific resistance is given by Ayrton and Perry as more than 10^{25}, but it is probably much higher in selected samples. The most serviceable kind of paraffin is the hardest obtainable, melting at a temperature of not less than 52 C. It is a good plan to remelt the commercial paraffin and keep it at a temperature of, say, 120 C. for an hour, stirring it carefully with a glass rod so that it does not get overheated; this helps to get rid of traces of water vapour.

Hard paraffin, when melted, has an enormous rate of expansion with temperature, so great, indeed, that care must be taken not to overfill the vessels in which it is to be heated. Castings can only be prepared by cooling the mould slowly from the bottom, keeping the rest of the mould warm, and adding-paraffin from time to time to make up for the contraction. The cooling is gradually allowed to spread up to the free surface.

The chief use of paraffin in the laboratory is in the construction of complicated connection boards, which are easily made by drilling holes in a slab of paraffin, half filling them with mercury, and using them as mercury cups.

Since paraffin is a great collector of dust, it should be screened by paper, otherwise the blocks require to be scraped at frequent intervals, which, of course, electrifies them considerably. This electrification is often difficult to remove without injuring the insulating power of the paraffin. A light touch with a clean Bunsen flame is the readiest method, and does not appear to reduce the insulation so much as might be expected. The safest way, however, is to leave the key covered by a clean cloth, which, however, must not touch the surface, for a sufficient time to allow of the charges getting away.

The paraffin often becomes electrified itself by the friction of the key contacts, so that in electrometer work it is often convenient to form the cups by lining them with a little roll of copper foil twisted up at the bottom. In this case the connecting wires should, of course, be copper. Small steel staples are convenient for fastening the collecting wires upon the paraffin; or, in the case where these wires have to be often removed and changed about, drawing-pins are very handy.

With mercury cups simply bored in paraffin great trouble will often be experienced in electrometer work, owing to a potential difference appearing between the cups — at all events when the contacts are inserted and however carefully this be done. A few drops of very pure alcohol poured in above the mercury often cures this defect. The surface of paraffin is by no means exempt from the defect of losing its insulating power when exposed to damp air, but it is not so sensitive as glass, nor does the insulating power fall so far.

Two useful appliances are figured.

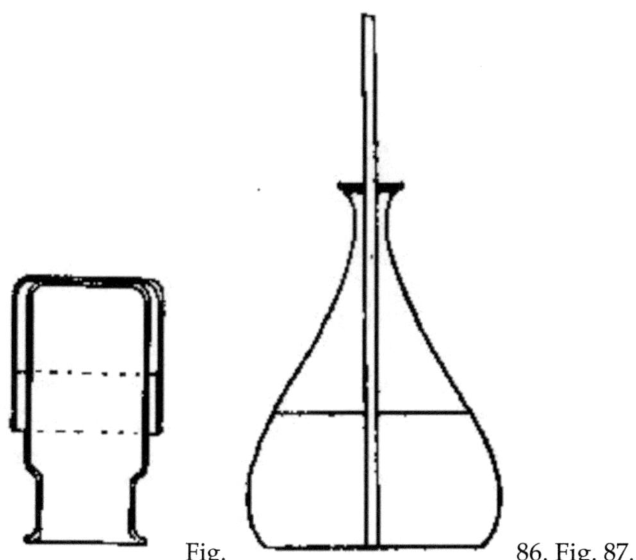

Fig. 86. Fig. 87.

One, in which paraffin appears as a cement, is an insulating stand made out of a bit of glass or ebonite tube cemented into an Erlenmeyer flask, having its neck protected from dust when out of use by a rubber washer, the other a "petticoat" insulator made by cementing a flint glass bottle into a glass dish with paraffin. In course of time the paraffin will be found to have separated from the glass. When this occurs the apparatus may be melted together again by placing it on the water bath for a few minutes.

113. Vaseline, Vaseline Oil, and Kerosene Oil. —

These, when dry, insulate almost, but not quite as well as solid paraffin. H. Koeller (*Wien Berichte*, 98, ii. 201, 1889; *Beibl. Wied. Ann.* 1890, p. 186), working with very small voltages, places the final(?) specific resistance of commercial petroleum, ether, and vaseline oil at about 2 X 10^{27} C.G.S. This is ten times higher than the value assigned to commercial benzene (C_6H_6), and a hundred times higher than the value for commercial toluene.

All these numbers mean little or nothing, but the petroleum and vaseline oil were the best fluid insulators examined by Koeller. By mixing vaseline with paraffin a soft wax may be made of any desired degree of softness, and by dissolving vaseline in kerosene an insulating liquid of any degree of viscidity may be obtained.

Hard paraffin may be softened somewhat by the addition of kerosene, and an apparently homogeneous mass cast from the mixture. It will be found, however, that in course of time the kerosene oozes out, unless present in very small quantity. Koeller has found (*loc. cit.*) that some samples of vaseline oil conducted "vollstaendig gut," but I have not come across such samples. Vaseline oil, however, is sold at a price much above its value for insulating purposes.

Kerosene oil is best obtained dry by drawing it directly from a new tin and exposing it to air as little as possible. Of course, it may be dried by chemical means and distillation, but this is usually (or always) unnecessary.

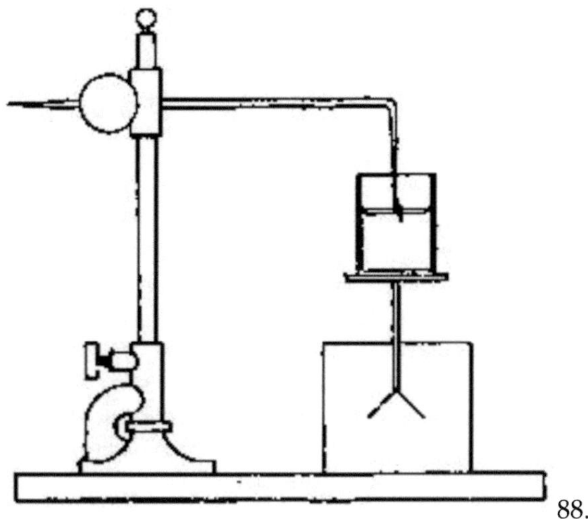

Fig 88.

There is some danger of kerosene containing minute traces of sulphuric acid, and it and other oils may be conveniently tested for insulation in the following manner. The quartz electroscope is taken, and the insulating rod heated in the blow-pipe. The electroscope will now insulate well enough to show no appreciable collapse of the leaves in one or two hours' time. Upon the plate of the electroscope is put a platinum or copper cylinder, and this is filled with kerosene (say) up to a fixed mark.

The electroscope is placed on a surface plate, or, at all events, on a sheet of plate glass, and a "scribing block" is placed along side it and the scriber adjusted to dip into the kerosene to any required depth. This is done by twisting a bit of wire round the scribing point and allowing it to project downwards. The point itself serves to give an idea of the height to which the vessel may be filled. The liquid is adjusted till its surface is in contact with the end of the scribing point, and the wire then projects into the liquid and forms an electrode of constant area of surface. The scribing block is put to earth. A charge is given to the electroscope, and the time required for a given degree of collapse of the leaves noted.

The kerosene is then removed and its place taken by vaseline or paraffin, known to insulate well as a standard for comparison. The experiment is then repeated, and the time noted for the same degree of collapse. This test, though of course rough, is generally quite sufficient for workshop purposes, and is easily applied. Moreover, it does not require correction for electrometer leakage, as generally happens when more elaborate appliances are used.

The actual resistance of insulating oils depends so much on the electrical intensity, on the duration of that intensity, and on the previous history of the oil as to the direction of the voltage to which it has been subjected — to say nothing of the effects of traces of moisture — that quantitative experiments are of no value unless they are extremely elaborate. I shall therefore only append the following numbers due to Bouty, *Ann. de Chemie et de Physique* (6), vol. xxvii. p. 62, 1892, in which the effect of the conductivity on the determination of the specific inductive capacity was properly allowed for: —

	Carbon Bisulphide.	Turpentine.	Benzene (C_6H_6) at 20 C.	Benzene at 60 C.
Specific inductive capacity	2.715	2.314	2.21	2.22
Specific resistance in ohms per cubic centimetre	1.5 x 10^{13},	1.75 x 10^{12}	1.56 x 10^{11}	7.9 x 10^{11}

[Footnote: Professor J. J. Thomson, and Newall (Phil. Proc. 1886) consider that carbon bisulphide showed traces of a "residual charge" effect; hence, until this point is cleared up, we must regard Bouty's value as corresponding only to a very short, but not indefinitely short, period of charge. On this point the paper must be consulted.

March 1897 — The writer has investigated this point by an independent method, but found no traces of "residual charge."]

Information as to the specific inductive capacity of a large number of oils may be found in a paper by Hopkinson, *Phil. Proc.* 1887, and in a paper by Quincke in Wiedemann's *Annalen*, 1883.

114. Imperfect Conductors. —

Under this heading may be grouped such things as wood, slate, marble, etc. — in fact, materials generally used for switchboard insulation. An examination of the insulating power of these substances has recently been made by B. O. Peirce (*Electrical Review*, 11th January 1895) with quite sufficient accuracy, having in view the impossibility of being certain beforehand as to the character of any particular sample. The tests were made by means of holes drilled in slabs of the material to be examined. These holes were three-eighths of an inch in diameter, and from five-eighths to three-quarters of an inch deep, and one inch apart, centre to centre. A voltage of about 15 volts was employed. The following general results were arrived at:-

(1) Heating in a paraffin bath always increases the resistance of wood, though only slightly if the wood be hard and dense.

(2) Frequent exhaustion and readmission of air above the surface of the paraffin always has a good effect in increasing the resistance of wood.

(3) When wood is once dry, impregnating it with paraffin tends to keep it dry.

(4) Red vulcanised fibre, like wood, absorbs paraffin, but it cannot be entirely waterproofed in this way.

(5) The resistance of wood with stream lines along the grain is twenty to fifty per cent lower than when the stream lines cross the grain.

(6) The "contact" resistance between slabs of wood pressed together is always very high.

The following table will sufficiently illustrate the results obtained. The stone was dried in the sun for three weeks in the summer (United States), and the wood is described as having been well seasoned: —

CURRENT WITH THE GRAIN

	Lowest Resistance between two Cups in Meg-ohms.	Highest Resistance between two Cups in Meg-ohms.	Lowest Specific Resistance in Meg-ohms.	Highest Specific Resistance in Meg-ohms.
Ash.	550	920	380	700
Cherry	1100	4000	2800	6000
Maho-gany	430	730	310	610
Oak	220	420	1050	2200
Pine.	330	630	360	1470
Hard pi-ne.	10	48	17	1050
Black walnut	1100	3000	320	2100
Red fib-re	2	4	3	60
Slate			184	280

| Soaps-
tone. | 330 | 500 |
| White
marble | 2000 | 8800 |

115. As to working the materials very little need be said.

Fibre is worked like wood, but has the disadvantage of rapidly taking the edge off the tools. In turning it, therefore, brass-turning tools, though leaving not quite such a perfect finish as wood-turning tools, last much longer, and really do well enough. Fibre will not bear heating much above 100 C. — at all events in paraffin. At 140 C. it becomes perfectly brittle. Its chief merit lies in its great strength. So far as insulation is concerned, Mr. Peirce's experiments show that it is far below most kinds of wood.

Slate. — This is a vastly more useful substance than it is generally credited with being. It is very easily worked at a slow speed, either on the shaping machine or on the lathe, with tools adjusted for cutting brass, and it keeps its figure, which is more than can be said for most materials. It forms a splendid base for instruments, especially when ground with a little emery by iron or glass grinders, fined with its own dust, and French polished in the ordinary way. Spools for coils of considerable radial dimension may be most conveniently made of slate instead of wood or brass, both because it keeps its shape, and because it insulates sufficiently well to stop eddy currents — at all events, sufficiently for ordinary practice. An appreciable advantage is that slate may be purchased at a reasonable rate in large slabs of any desired thickness. It is generally cut in the laboratory by means of an old cross-cut saw, but it does not do much damage to a hard hack saw such as is used for iron or brass.

Marble. — According to Holtzapffell, marble may be easily turned by means of simple pointed tools of good steel tempered to a straw colour. The cutting point is ground on both edges like a wood-turning tool, and held up to the work "at an angle of twenty or thirty degrees" (?with the horizontal). The marble is cut wet to save the tool. As soon as the point gets, by grinding, to be about one-eighth of an inch broad it must either be drawn down or reground; a flat tool will not turn marble at all.

A convenient saw for marble is easily made on the principle of the frame saw. A bit of hoop iron forms a convenient blade, and is sharpened by being hammered into notches along one edge, using the sharp end of a hammer head. The saw is liberally supplied with sand and water — or emery and water, where economy of time is an object. The sawing of marble is thus really a grinding process, but it goes on rapidly. Marble is ground very easily with sand and water; it is fined with emery and polished with putty powder, which, I understand, is best used with water on cloth or felt. As grinding processes have already been fully described, there is no need to go into them here. I have no personal knowledge of polishing marble.

116. *Conductors.* —

The properties of conductors, more particularly of metals, have been so frequently examined, that the literature of the subject is appallingly heavy. In what follows I have endeavoured to keep clear of what might properly appear in a treatise on electricity on the one hand, and in a wiring table on the other. The most important work on the subject of the experimental resistance properties of metals has been done by Matthieson, *Phil. Trans.* 1860 and 1862, and *British Association Reports* (1864); Callender, *Phil. Trans.* vol. clxxiii.; Callender and Griffiths, *Phil. Trans.* vol. clxxxii.; *The Committee of the British Association on Electrical Standards from 1862 to Present Time;* Dewar and Fleming, *Phil. Mag.* vol. xxxvi. (1893);

Klemencic, *Wiener Sitzungsberichte* (Denkschrift), 1888, vol. xcvii. p. 838; Feussner and St. Lindeck, *Zeitsch. fuer Inst. 'Kunde*, ix. 1889, p. 233, and *B. A. Reports*, 1892, p. 139. Of these, Matthieson, and Dewar and Fleming treat of resistance generally, the latter particularly at low temperatures.

[Footnote: The following is a list of Dr. Matthieson's chief papers on the subject of the electrical resistance of metals and alloys: *Phil. Mag.* xvi. 1858, pp. 219-223; *Phil. Trans.* 1858, pp. 383-388 *Phil. Trans.* 1860, pp. 161-176; *Phil. Trans.* 1862, pp. 1-27 *Phil. Mag.* xxi. (1861),

pp. 107-115; *Phil. Mag.* xxiii. (1862), pp. 171-179; *Electrician*, iv. 1863, pp. 285-296; *British Association Reports*, 1863, p. 351.]

Matthieson, and Matthieson and Hockin, Klemencic, Feussner, and St. Lindeck deal with the choice of metals for resistance standards. Callender's, and Callender and Griffiths' work is devoted to the study of platinum for thermometric purposes.

The bibliography referring to special points will be given later. The simplest way of exhibiting the relative resistances of metals is by means of a diagram published by Dewar and Fleming (*loc. cit.*), which is reproduced on a suitable scale on the opposite page. For very accurate work, in which corrections for the volumes occupied by the metals at different temperatures are of importance, the reader is referred to the discussion in the original paper, which will be found most pleasant reading. From this diagram both the specific resistance and the temperature coefficient may be deduced with sufficient accuracy for workshop purposes. In interpreting the diagram the following notes will be of assistance. The diagram is drawn to a scale of so-called "platinum temperatures" — that is to say, let R_0, R_{100}, R_t be the resistances of pure platinum at 0, 100, and t C. respectively, then the platinum temperature p_t is defined as

$$p_t = 100 \times (R_t - R_0)/(R_{100} - R_0)$$

This amounts to making the temperature scale such that the temperature at any point is proportional to the resistance of platinum at that point. Consequently on a resistance temperature diagram the straight line showing the relation between platinum resistance and platinum temperature will "run out" at the platinum absolute zero, which coincides more or less with the thermodynamic absolute zero, and also with the "perfect gas" absolute zero. Platinum temperatures may be taken for workshop purposes over ordinary ranges as almost coinciding with air thermometer temperatures. The metals used by Professors Dewar and Fleming were, with some exceptions, not absolutely pure, but in general represent the best that can be got by the most refined process of practical metallurgy. We may note further that the specific resistance is only correct for a temperature of about 15 C., since no correction for the expansion or contraction of material has been applied.

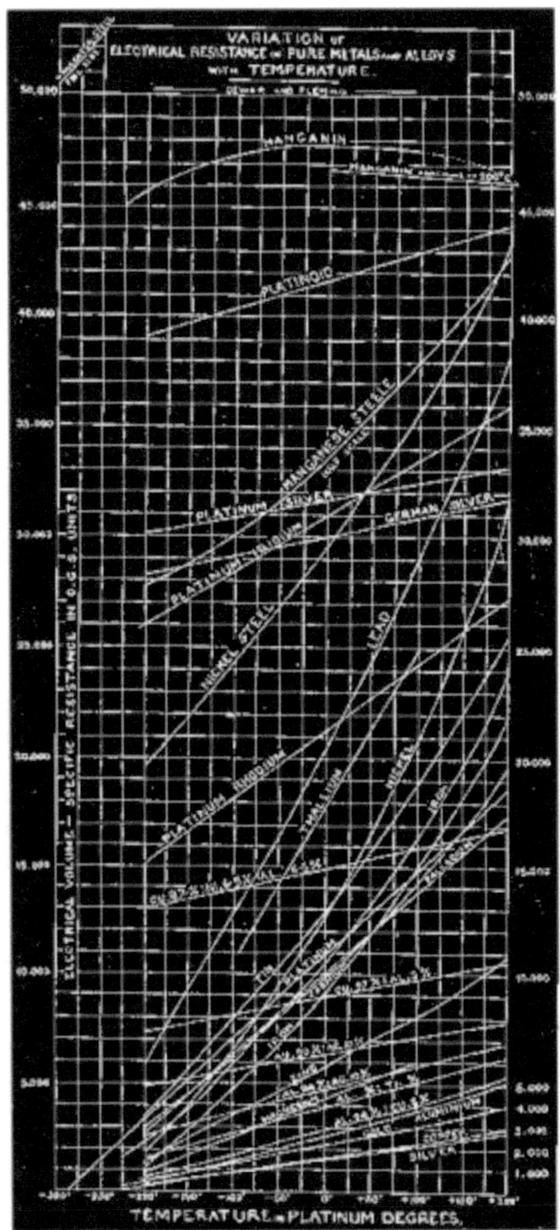

VARIATION of
ELECTRICAL RESISTANCE of PURE METALS and ALLOYS
with TEMPERATURE.
MANGANIN
MANGANIN annealed at 300°C.
PLATINOID
MANGANESE STEEL
PLATINUM SILVER
GERMAN SILVER
PLATINUM IRIDIUM
NICKEL STEEL
LEAD
PLATINUM RHODIUM
PLATINUM
NICKEL
IRON
TIN
PLATINUM
ELECTRICAL VOLUME — SPECIFIC RESISTANCE IN C.G.S. UNITS
ALUMINIUM
GOLD
COPPER
SILVER
TEMPERATURE in PLATINUM DEGREES.

The following notes on alloys suitable for resistance coils will probably be found sufficient.

117. *Platinoid.* —

This substance, discovered by Martino and described by Bottomley (*Phil. Proc. Roy. Soc.* 1885), is an alloy of nickel, zinc, copper, and 1 per cent to 2 per cent of tungsten, but I have not been able to obtain an analysis of its exact composition. It appears to be difficult to get the tungsten to alloy, and it has to be added to part of the copper as phosphide of tungsten, in considerably greater quantity than is finally required. The nickel is added to part of the copper and the phosphide of tungsten, then the zinc, and then the rest of the copper. The alloy requires to be remelted several times, and a good deal of tungsten is lost by oxidation.

The alloy is of a fine white colour, and is very little affected by air — in fact, it is to some extent untarnishable. The specific resistance will be seen to be about one and a half times greater than that of German silver, and the temperature coefficient is about 0.021 per cent per degree C. (i.e. about nineteen times less than copper, and half that of German silver). To all intents and purposes it may be regarded as German silver with 1 per cent to 2 per cent of tungsten. It does not appear to have been particularly examined for secular changes of resistance.

118. *German Silver.* — This material has been exhaustively examined of late years by Klemencic and by Feussner and St. Lindeck. Everybody agrees that German silver, as ordinarily used for resistances, and composed of copper four parts, zinc two parts, nickel one part, is very ill-fitted for the purpose of making resistance standards. This is due

(1) to its experiencing a considerable increase in resistance on winding. Feussner and St. Lindeck found an increase of 1 per cent when German silver was wound on a core of ten wire diameters.

(2) To the fact that the change goes on, though with gradually decreasing rate, for months or years;

(3) to the fact that the resistance is permanently changed (increased) by heating to 40 C. or over. By "artificially ageing" coils of

German silver by heating to 150 C., say for five or six hours, its permanency is greatly improved, and it becomes fit for ordinary resistance coils where changes of, say, 1/5000 do not matter.

It is a remarkable property of all nickel alloys containing zinc that their specific resistance is permanently increased by heating, whereas alloys which do not contain zinc suffer a change in the opposite direction. The manufacturers of German silver appear to take very little care as to the uniformity of the product put on the market; some so-called German silver is distinctly yellow, while other samples are bright and white.

It is noted by Price (*Measurements of Electrical Resistance*, p. 24) that German silver wire is apt to exhibit great differences of resistance within quite short lengths. This has been my own experience as well, and is a great drawback to the use of German silver in the laboratory, for it makes it useless to measure off definite lengths of wire with a view to obtaining an approximate resistance. In England German silver coils are generally soaked in melted hard paraffin. In Germany, at all events at the Charlottenburg Institute, according to St. Lindeck — coils are shellac-varnished and baked. In any case it appears to be essential to thoroughly protect the metal against atmospheric influence.

119. Platinum Silver. —

In the opinion of Matthieson and of Klemencic the 10 per cent silver, 90 per cent platinum alloy is the one most suitable for resistance standards. At all events, it has stood the test of time, for, with the following exceptions, all the British Association coils constructed of it from 1867 to the present day have continued to agree well together. The exceptions were three one-ohm coils, which permanently increased between 1888 and 1890, probably through some straining when immersed in ice. One coil changed by 0.0006 in 1 between the years 1867 and 1891. According to Klemencic, absolute permanency is not to be expected even from this alloy.

Its recommendation as a standard depends on its chemical inertness, its small temperature coefficient (0.00027 per degree), and its small thermo-voltage against copper, as the following table (taken from Klemencic) will show: —

Thermo-voltages in Micro-volts per degree against Copper over the Range 0 to 17 C.

Platinum iridium	7.14 micro-volts per degree C.
Platinum silver	6.62 micro-volts per degree C.
Nickelin .	28.5 micro-volts per degree C.
German silver	10.43 micro-volts per degree C.
Manganin (St. Lindeck)	1.5 micro-volts per degree C.

Mechanically, the platinum silver is weak, and is greatly affected as to its resistance by mechanical strains — in fact, Klemencic considers it the worst substance he examined from this point of view — a conclusion rather borne out by Mr. Glazebrook's experience with the British Association standards already referred to (*B. A. Reports*, 1891 and 1892).

Taking everything into account, it will probably be well to construct standards either with oil insulation only, or to bake the coils in shellac before testing, instead of soaking in paraffin. Fig. 89 illustrates a form of an oil immersed standard which is in use in my laboratory, and through which a considerable current may be passed. The oil is stirred by means of a screw propeller.

Fig.

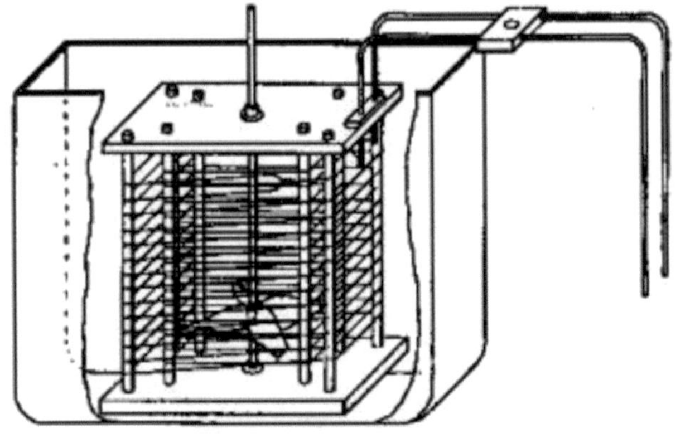

89.

Fig. 89 represents a standard resistance for making Clerk cell comparisons by the silver voltameter method. The framework on which the coils are wound consists of a base and top of slate. The pillars are of flint glass tube surrounding brass bolts, and cemented to the latter by raw shellac. Grooves are cut in the glass sleeves to hold the wires well apart. These grooves were cut by means of a file working with kerosene lubrication. A screw stirrer is provided, and the whole apparatus is immersed in kerosene in the glass box of a storage cell. The apparatus is aged to begin with by heating to a temperature a good deal higher than any temperature it is expected to reach in actual work. After this the rigidity of the frame is intended to prevent any further straining of the wire. The apparatus as figured is not intended to be cooled to 0 C., so that it is put in as large a box as possible to gain the advantage of having a large volume of liquid. The top and bottom slates measure seven inches by seven inches, and the distance between them is seven inches. The inner coil is wound on first, and the loop which constitutes the end of the winding is brought up to a suitable position for adjustment. The insulation of the heavy copper connectors is by means of ebonite.

120. Platinum Iridium. —

Platinum 90 per cent, iridium 10 per cent. This material was prepared in some quantity at the cost of the French Government, and distributed for test about 1886. Klemencic got some of it as representing Austria, and found it behaved very like the platinum silver alloy just discussed. The temperature coefficient is, however, higher than for platinum silver (0.00126 as against 0.00027). The mechanical properties of the alloy are, however, much better than those of the silver alloy; and in view of the experience with B. A. standards above quoted, it remains an open question whether, on the whole, it would not be the better material for standards, in spite of its higher price. Improvements in absolute measurements of resistance, however, may render primary standards superfluous.

121. *Manganin.* —

Discovered by Weston — at all events as to its application to resistance coils. A glance at the diagram will exhibit its unique properties, on account of which it has been adopted by the Physikalisch Technischen Reichsanstalt for resistance standards. The composition of the alloy is copper 84 per cent, manganese 12 per cent, nickel 4 per cent., and it is described as of a steel-gray colour. Unfortunately it is apt to oxidise in the air, or rather the manganese it contains does so, so that it wants a very perfect protection against the atmosphere.

Like German silver, manganin changes in resistance on winding, and coils made of it require to be artificially aged by heating to 150 for five hours before final adjustment. The annealing cannot be carried out in air, owing to the tendency to oxidation. The method adopted by St. Lindeck (at all events up to 1892) is to treat the coil with thick alcoholic shellac varnish till the insulation is thoroughly saturated, and then to bake the coil as described. The baking not only anneals the wire, but reduces the shellac to a hard and highly insulating mass.

Whether stresses of sufficient magnitude to produce serious mechanical effects can be set up by unequal expansion of wire and shellac during heating and cooling is not yet known, but so far as tested (and it must be presumed that the Reichsanstalt tests are thorough) no difficulty seems to have been met with. In course of time, however, probably the best shellac coating will crack, and then adieu to the permanency of the coil! This might, of course, be obviated by keeping the coil in kerosene, which has no action on shellac, but which decomposes somewhat itself.

The method of treatment above described suffices to render coils of manganin constant for at least a year (in 1892 the tests had only been made for this time) within a few thousands per cent. Manganin can be obtained in sheets, and from this material standards of 10^{-2}, 10^{-3}, and 10^{-4} ohms are made by soldering strips between stout copper bars, and these are adjusted by gradually increasing their resistance by boring small holes through them. The solder employed is said to be "silver."

Mr. Griffiths (*Phil. Trans.* vol. clxxxiv. [1893], A, p. 390) has had some experience with manganin carrying comparatively heavy currents, under which circumstances its resistance when immersed in water was found to rise in spite of the varnish which coated it. Other experiments in which the manganin wire was immersed in paraffin oil did not exhibit this effect, though stronger currents were passed.

On the whole, manganin appears to be the best material for coil boxes and "secondary" resistance standards. Whether it is fit to rank with the platinum alloys as regards permanency must be treated as an open question.

122. *Other Alloys.* —

The following tables, taken from the work of Feussner and St. Lindeck, *Zeitschrift fuer Instrumenten Kunde*, 1889, vol. ix. p. 233, together with the following notes, will suffice.

123. *Nickelin.* —

This is only German silver with a little less zinc, a little more nickel, and traces of cobalt and manganese. It behaves like German silver, but is an improvement on the latter in that all the faults of German silver appear upon a reduced scale in nickelin.

124. Patent Nickel. —

Practically a copper nickel alloy, used to some extent by Siemens and Halske. It stands pretty well in the same relation to nickelin as the latter does to German silver. After annealing as for manganin it can be made into serviceable standards which do not change more than a few thousandths per cent. I have not come across a statement of its thermo-voltage against copper.

125. *Constantin.* —

Another nickel copper alloy containing 50 per cent of each constituent. It appears to be a serviceable substance, having a temperature coefficient of 0.003 per cent per degree only, but an exceedingly high thermo-voltage, viz. 40 micro-volts per degree against copper.

1	2	3	4	5	6	7.	8
German Silver.	Nickelin made by Obermaier		Rhe-o-tane.	Patent Nickel		Manga-nese Copper.	ck… Ma… ne… Co…
	Diameter 1.0 mm	Diameter 0.1 mm		Diameter 0.6 mm	Diameter 1.0 mm		
60.16	61.63	54.57	53.28	74.41	74.71	70	
25.37	19.67	20.44	16.89	0.23	0.52	…	
	…	…	…	…	trace	…	
14.0	18.4	24.4	25.3	25.1	24.1	…	

	3	6	8	1	0	4	
n	0.30	0.24	0.64	4.46	0.42	0.70	...
	trace	0.19	...	...	trace	trace	...
ng e	trace	0.18	0.27	0.37	0.13	0.17	30
	99.86	100.37	100.40	100.31	100.24	100.24	...
-	30.0	33.2	44.8	52.5	34.2	32.8	100.6
p re nt	0.00036	0.00030	0.00033	0.00041	0.00019	0.00021	0.0000043

The specific resistance is in "microhms, i.e. 10^{-6} ohms per cubic centimetre, and the temperature coefficient in degrees centigrade.

126. Nickel Manganese Copper. —

I can find no other reference with regard to this alloy mentioned by Lindeck. Nicholls, however (Silliman's Journal [3], 39, 171, 1890), gives some particulars of alloys of copper and ferromanganese. The following table is taken from Wiedemann's *Beiblatter* (abstract of Nicholl's paper, 1890, p.

811). All these alloys appear to require annealing at a red heat before their resistances are anything like constant.

Let x be percentage of copper, then $100 - x$ is percentage of "ferromanganese."

Values of x.	100	99.26	91.88	86.98	80.4	70.65
Specific resistance with respect to copper (? pure)	1	1.19	11.28	20.4	27.5	45.1
Temperature coefficient per degree x 10^6 (hard)	3202	2167	138	16	22	-24
Ditto (soft)			184	80	66	21

If nickel is added, alloys of much the same character are obtained, some with negative temperature coefficients — for instance, one containing 52.51 per cent copper, 31.27 per cent ferromanganese, and 16.22 nickel.

A detailed account of several alloys will be found in a paper by Griffiths (Phil. Trans. 1894, p. 390), but as the constants were determined to a higher order of accuracy than the composition of the material — or, at all events, to a higher degree of accuracy than that to which the materials can be reproduced — there is no advantage in quoting them here.

CHAPTER IV

127. Electroplating. —

This is an art which is usually deemed worthy of a treatise to itself, but for ordinary laboratory purposes it is a very simple matter — so simple, indeed, that the multiplicity of receipts as given in treatises are rather a source of embarrassment than otherwise.

The fundamental principles of the art are:-

(1) Dirty work cannot be electroplated.

(2) Electroplated surfaces may be rougher, but will not be smoother than the original unplated surface.

(3) The art of electroplating being in advance of the science, it is necessary to be careful as to carrying out instructions in detail. This particularly applies to the conditions which determine whether a metallic deposit shall come down in a reguline or in a crystalline manner.

128. The Dipping Bath. —

An acid dipping bath is one of the most useful adjuncts to the laboratory, not only for cleansing metals for electroplating, but for cleaning up apparatus made out of bits of brass tube and sheet, and particularly for quickly cleaning binding screws, etc., where it is necessary to ensure good electrical contact.

The cheapest and most satisfactory way in the end is to make up two or three rather large baths to begin with. The glass boxes of storage batteries do very nicely for the purpose, and being generally ground pretty flat at the top, they may be covered by sheets of patent plate glass, and thus preserved from the action of the air.

First Bath. — A 30 or 40 per cent solution of commercial caustic soda. Objects may be cleansed from grease in this bath by heating them as hot as is consistent with individual circumstances, and plunging them into it.

It is a considerable advantage to begin by removing grease from articles subsequently to be dipped in an acid bath, both because it saves time and acid, and because more uniform results are obtainable when this is done than when it is omitted. It is a great advantage to have the caustic soda solution hot. This is always done in factories where nickel-plating is carried on, but it is inconvenient in the laboratory. The articles after dipping in the alkali are swilled with water, and may even be scrubbed with a brush, so as to remove greasy matters that have been softened but not entirely removed.

Acid Bath. — A convenient bath for laboratory purposes is made by mixing two volumes of strong commercial nitric acid with one of strong sulphuric acid in a cell measuring, say, 12 X 10 X 15 inches.

Copper or brass articles are dipped in this bath for a few seconds, then rinsed with water, then dipped again for a second or two, or until they appear equally white all over, and then withdrawn as rapidly as possible and plunged into a large quantity of clean water. Care must be taken to transfer the articles from the bath to the water as quickly as possible, for if time be allowed for gas to be evolved, the surfaces become mat instead of bright.

In order to save acid it is advisable to make up a third bath, using those odds and ends of acids which gradually accumulate in the laboratory. Sulphuric acid from the balance cases, for instance, mixed with its own volume of commercial nitric acid, does very well.

The objects to be dipped receive a preliminary cleansing by a dip in this bath, the strong bath being reserved for the final dip. Sheet brass and drawn tube, as it comes from the makers, possesses a really fine surface, though this is generally obscured by grease and oxide. Work executed in these materials, cleaned in alkali, and dipped in really strong acid, will be found to present a much better appearance than work which has been filed, unless the latter be afterwards elaborately polished.

On no account must paraffin be allowed to get into any of the baths. When the final bath gets weak it must be relegated to a subordinate position and a new bath set up. A weak acid bath leaves an ugly mottled surface on brass work.

129. A metallic surface which it is intended to electroplate must, as has been mentioned, be scrupulously clean. If the metal is not too

valuable or delicate, cleaning by dipping is easy and effectual. The following notes will be found to apply to special cases which often occur.

(1) *Silver Surfaces intended to be gilt.* — These are first washed clean with soap and hot water, and polished with whitening. They are then dipped for a moment in a boiling solution of potassium cyanide. A 20 per cent solution of common commercial cyanide does well, but the exact strength is quite immaterial. The cyanide is washed away in a large volume of soft water, and the articles are kept under water till they are scratch-brushed.

Mat surfaces are readily produced on standard silver by dipping in hot strong sulphuric acid. The appearance of new silver coins, which is familiar to everybody, is obtained by this process.

(2) *Finely turned and finished Brass Work.* — If it is intended to nickel-plate such work, and if it is desirable to obtain brightly polished nickel surfaces, the work must be perfectly polished to begin with. Full details as to polishing may be found in workshop books or treatises on watch-making. It will suffice here to say that the brass work is first smoothed by the application of successive grades of emery and oil, or by very fine "dead" smooth files covered with chalk. Polishing is carried out by means of rotten stone and oil applied on leather.

In polishing turned work care must be taken to move the file, emery, or rotten stone to and fro over the work with great regularity, or the surface will end by looking scratchy and irregular. The first process of cleaning is, of course, to remove grease, and this is accomplished best by dipping in a bath of strong hot caustic soda solution, and less perfectly by heating the work and dipping it in the cold caustic soda bath.

During this process a certain amount of chemical action often occurs leading to the brass surface exhibiting some discoloration. The best way of remedying this is to dip the brass into a hot bath of cyanide of potassium solution. If it is inconvenient to employ hot baths or to heat the brass work, good results may be obtained by rubbing the articles over with a large rough cork plentifully lubricated with a strong solution of an alkali.

If the surfaces are very soiled or dirty, a paste of alkali and fine slaked lime may be applied on a cork rubber, and this in my experience has always been most effective and satisfactory in every way, except that it is difficult to get into crevices. If the alkali stains the work, a little cyanide of potassium may be rubbed over the surface in a similar manner.

Brass work treated by either of these methods is to be washed in clean water till the alkali is entirely removed, and may then be nickel-plated without any preliminary scratch-brushing. The treatment in hot baths of alkali and cyanide is the method generally employed in American factories as a preliminary to the nickelling of small brass work for sewing machines, etc.

(3) *Copper* either for use as the kathode in electrolysis calibration experiments or otherwise is most conveniently prepared by dipping in the acid bath, rinsing quickly in cold water, scratch-brushing under cold water, and transferring at once to the plating bath. In the case where the copper plates require to be weighed they are dipped into very hot distilled water after scratch-brushing, and then dried at once by means of a clean glass cloth.

(4) *Aluminium* (which, however, does not readily lend itself to plating operations *[Footnote:* This difficulty has now been overcome. See note, section 138.*]*) is best treated by alkali rubbed on with a cork, or by a hot alkaline carbonate where rubbing is inexpedient. The clean aluminium is scratch-brushed under water, and at once transferred to the plating bath.

(5) *Iron for Nickel-plating.* — According to Dr. Gore (Electra-metallurgy, p. 319) the best bath for cleaning iron is made as follows: "One gallon of water and one pound of sulphuric acid are mixed with one or two ounces of zinc (which of course dissolves); to this is added half a pound of nitric acid." The writer has been accustomed to clean iron by mechanical means, to deprive it of grease by caustic alkali, and to finish it off by, means of a hard scratch brush. This process has always worked satisfactorily.

(6) *Articles soldered with soft solder containing lead and tin* do not readily lend themselves to electrolytic processes, the solder generally becoming black and refusing to be coated with the electro-deposit. Moreover, if soldered articles are boiled for any length of

time in caustic alkali during the preliminary cleansing, enough tin will dissolve to form a solution of stannate of potash or soda — strong enough to deposit tin on brass or copper. A method of coppering soldered articles will be described later on.

130. Scratch-brushing. —

This process is generally indispensable, and to its omission is to be traced most laboratory failures in electroplating. Scratch-brushes may be bought at those interesting shops where "watchmakers' supplies" are sold. It will be well, therefore, to purchase a selection of scratch brushes, for they are made to suit particular kinds of work. They are all made of brass wire, and vary both in hardness and in the fineness of the wire. The simplest kind of scratch brush consists merely of a bundle of wires bound up tightly by another wire, and somewhat "frizzed" out at the ends (Fig. 90). A more useful kind is made just like a rotating brush, and has to be mounted on a lathe (Fig. 91).

Fig. 90. Fig. 91.

The scratch brush is generally, if not always, applied wet; the lubricant generally recommended is stale beer, but this may be replaced by water containing a small quantity of glue, or any other form of gelatine in solution — a mere trace (say .1 per cent) is quite sufficient. Very fair results may be got by using either pure or soapy water. The rotating brushes require to be mounted on a lathe, and may be run at the same speed as would be employed for turning wooden objects of the same dimensions.

Since the brush has to be kept wet by allowing water or its equivalent to drip upon it, it is usual to make a tin trough over which the brush can revolve, and to further protect this by a tin hood to keep the liquid from being thrown all over the room. In many works the brush is arranged to lie partly in the liquid, and this does very well if the hood is effective.

There is a superstition that electro-deposits stick better to scratch-brushed surfaces than to surfaces which have not been so treated, and consequently it is usual to scratch-brush surfaces before electro-deposit. However this may be, there is no doubt that adherence and solidity are promoted by frequent scratch-brushing during the process of depositing metal, especially when the latter tends to come down in a spongy manner.

Gilt surfaces — if the gilding is at all heavy — are generally dull yellow, or even brown, when they come from the bath, and require the scratch brush to cause the gold to brighten, an office which it performs in a quite striking manner. The same remark applies to silvered surfaces, which generally leave the bath a dead white — at all events if the deposit is thick, and if ordinary solutions are employed. In either case the touch of the scratch brush is magical.

131. Burnishing. —

Burnishers of steel, agate, or bloodstone can be bought at the shops where scratch brushes are sold, and are used to produce the same brightening effect as can be got by scratch-brushing. The same solutions are employed, but rather stronger, and the burnisher is swept over the surface so as to compress the deposited metal. Burnishing is rather an art, but when well done gives a harder and more brilliant (because smoother) surface than the scratch brush. On the whole, steel burnishers are the most convenient if in constant use.

If the burnishing tools have to lie about, steel is apt to rust, unless carefully protected by being plunged in quicklime or thickly smeared with vaseline, and the least speck of rust is fatal to a burnisher. In any case the steel requires to be occasionally repolished by rouge and water on a bit of cloth or felt. The process of burnishing is necessarily somewhat slow and tedious, and as a rule is not worth troubling about except in cases where great permanence is required.

The burnisher is moved over the work somewhat like a pencil with considerable pressure, and care is taken to make the strokes as uniform in direction as possible; otherwise the surface looks non-uniform, and has to be further polished by tripoli, whitening, etc., before it is presentable.

132. Silver-plating. —

The most convenient solution for general purposes is an 8 to 10 per cent solution of the double cyanide of silver and potassium together with 1 or 2 per cent of "free" potassium cyanide. Great latitude is permissible in the strength of solution and density of current. As commercial cyanide of potassium generally contains an unknown percentage of other salts, which, however, do not interfere with its value for the purpose of silver-plating, the simplest procedure is as follows.

For every 100 c.c. of plating solution about 7 grms. of dry crystallised silver nitrate are required. The equivalent amount of potassium cyanide (if dry and pure) is 5.2 grms., but commercial cyanide may contain from 50 per cent upwards to 96 per cent in the best fused cyanide made from ferrocyanide only. An approximate idea of the cyanide content can be obtained from the dealers when the salt is purchased, and this is all that is required.

A quantity slightly in excess of the computed amount of cyanide is dissolved in distilled water, and this is cautiously added to the solution of the silver nitrate till precipitation is just complete. The supernatant liquors are then drained away, and the precipitate dissolved by adding a sufficiency of the remaining cyanide; this process is assisted by warming and stirring.

An allowance of about one-tenth of the whole cyanide employed may be added to form "free" cyanide, and the solution made up to the strength named. It is advisable to begin with the cyanide in a moderately strong solution, for the sake of ease in dissolving the precipitate.

This solution will deposit silver upon articles of copper or brass immersed in it even without the battery, but the coat will be thin. The solution is used cold, with a current density of about 10 to 20 amp航s per square foot. The articles to be silvered are scratch-brushed, washed, and electroplated, till they begin to look undesirably rough. They are then taken out of the bath, rebrushed, and the process continued till a sufficiency of silver is deposited. Four grammes weight of silver (nearly) is deposited per amp航 hour. It is best to use a fine silver anode, so that the solution, does not get contaminated by copper.

In most factories it is usual to "quicken" the objects to be silvered before placing them in the electrolysis vats, because the deposit is said to adhere better in consequence of this treatment. I have never found it any improvement for laboratory purposes, but it is easy to do. A dilute (say 2 per cent) solution of cyanide of mercury is required containing a little free cyanide. The objects to be "quickened" are scratch-brushed and dipped into the cyanide of mercury solution till they are uniformly white; it is generally agreed that the less the mercury deposited the better, so long as a perfect coating is obtained. The objects are rinsed after quickening, and put in the depositing bath at once.

The mat surface of silver obtained by electrolysis of the cyanide is very beautiful — one of the most beautiful things in nature — shining with incomparable crystalline whiteness. So delicate is it, however, for so great is the surface it exposes, that it is generally rapidly deteriorated by exposure to the air. It may be protected to some extent by lacquering with pale lacquer, but it loses some of its brilliancy and purity in the process. The deposit is generally scratch-brushed or burnished down to a regular reflecting surface.

133. Cold Silvering. —

A thin but brilliant coat of silver may be readily applied to small articles of brass or copper in the following way. A saturated solution of sodium sulphite (neutral) is prepared, and into this a 10 per cent solution of nitrate of silver is poured so long as the precipitate formed is redissolved. A good deal of silver may be got into solution in this way. Articles to be silvered need only to be cleaned, brushed, and dipped in this solution till a coat of the required thickness is obtained.

I must admit, however, that the coating thus laid on does not appear to be so permanent as one deposited by simple immersion from the cyanide solution, even though it is thicker. The cyanide plating solution will itself give a good coat of silver if it is used boiling, and if a little potassium cyanide be added.

For purposes of instrument construction, however, a thin coat of silver is seldom to be recommended, on account of its liability to tarnish and its rapid destruction when any attempt is made to re-polish it. For these reasons, nickel or gold plating is much to be preferred.

134. *Gilding.* —

This art deserves to be much more widely practised than is usual in laboratories. Regarded as a means of preserving brass, copper, or steel, it is not appreciably more "time robbing" than lacquering, and gives infinitely better results. Moreover, it is not much more expensive. Strange as it may seem, the costliness of gilding seldom lies in the value of the gold deposited; the chief cost is in the chemicals employed to clean the work, and in interest on the not inconsiderable outlay on the solution and anode.

The easiest metal to gild is silver, and it is not unusual to give base metals a thin coating of silver or copper, or both, one after the other, before gilding, in order to secure uniformity. To illustrate the virtue of a thin layer of gold, I will mention the following experiment. About three years ago I learned for the first time that to "clean" the silver used in a small household required at least an hour's labour *per diem*. I further ascertained that most of this time is spent on the polishing part of the process.

As this seemed a waste of labour, I decided to try the effect of gilding. In order to give the proposal a fair trial I gilt the following articles: half a dozen table spoons and forks, a dozen dessert forks and spoons, and a dozen tea spoons. These were all common electroplated ware. They were weighed before and after gilding, and it was with difficulty that the increase of weight was detected, even though a fine bullion balance was employed. On calculating back to money, it appeared that the value of the gold deposited was about threepence. Assuming that an equal weight of silver had been accidentally dissolved by the free cyanide during the plating — which is unlikely — the total amount of gold deposited would be worth, say, sixpence.

After three years' continuous use the gilding is still perfect, except at the points on which the spoons and forks rest, where it is certainly rather shabby. Meanwhile the "gold" plate only requires to be washed with hot water and soap to keep it in perfect order, a much more cleanly and expeditious process than that of silver cleaning.

135. Preparing Surfaces for Gilding. —

Ordinary brass work — rough or smooth — may for purposes of preservation be dipped, scratch-brushed, and gilt at once. Seven years ago the writer gilt the inside of the head of a copper water still, and simply scratch-brushed it; it is to-day in as good order as when it was first done. If it is intended to gild work from the first, with the view of making an exceptionally fine job of it, "gilding metal," i.e. brass containing one to one and a quarter ounces of zinc to the pound of copper may be specified. From its costliness, however, this is only desirable for small work.

Iron and steel are generally given a preliminary coating of copper, but this may be dispensed with though with no advantage — by using a particular process of gilding.

Base metals, zinc, pewter, lead, etc., are first coppered in a cyanide of copper solution, as will be described under the head of Copper-plating. If it is intended to gild soldered articles, the preliminary coating of copper is essential.

The most convenient vessel for holding a gilding solution is undoubtedly one formed of enamelled iron. Particularly useful are the buckets and "billies" (i.e. cylindrical cans) made of this material. These vessels may be heated without any fear of a smash, and do not appear to be appreciably affected by gilding solutions — at all events during several days or weeks. The avoidance of all risk of breakage when twenty or thirty pounds' worth of solution is in question is a matter of importance.

Under no circumstances is it desirable to use anything but the purest gold and best fused cyanide (called "gold" cyanide) in the preparation of the solutions. The appearance of a pure gold deposit is far richer than of one containing silver, and its resistance to the atmosphere is perfect; moreover, in chemico-physical processes one has the satisfaction of knowing what one is dealing with.

136. Gilding Solutions. —

The strength of solution necessary for gilding brass, copper, and silver is not very material. About one to two pounds of "gold" potassium cyanide (? 96 per cent KCN) per gallon does very well. The gold is best introduced by electrolysing from a large to a small gold electrode. One purchases a plate of pure gold either from the mint or from reliable metallurgists (say Messrs. Johnson and Matthey of London), and from this electrodes are cut.

The relative areas of the electrodes do not really much matter. I have used an anode of four times the area of the cathode. The solution is preferably heated to a temperature of about 50 C., and a strong current is sent through it, say twenty amperes to the square foot of anode. The electrodes must be suspended below the surface of the solution by means of platinum wires. If the gold plates are only partly immersed, they dissolve much more rapidly where they cut the surface, possibly on account of the effect of convection currents, though so far as the writer is aware no proper explanation has yet been given.

After a time gold begins to be deposited on the cathode in a powdery form, for which reason it is a good plan to begin by wrapping the latter in filter paper. The process has gone on for a sufficient time when a clean bit of platinum foil immersed in the place of the cathode becomes properly gilt at a current density of about ten amperes per square foot.

The powdery gold deposited on the cathode while preparing the solution can be scraped off and melted for further use, or the whole cathode may now be used as an anode. The platinum foil testing cathode may also be "stripped" by making it an anode, and is for this reason preferable to German silver or copper, which would contaminate the solution while the "stripping" process was in progress.

For general purposes a current density of say ten to fifteen amperes per square foot may be used, but this may be considerably varied, so long as the upper limit is not greatly overpassed. During gold-plating there is a considerable advantage in keeping the electrodes moving or the solution stirred.

After immersing the cleaned and scratch-brushed articles, depositing may go on for about three minutes, after which they are removed from the bath and examined, in order to detect any want of uniformity in the deposit.

The articles should be entirely immersed; if this is not done, irregularity is apt to appear at the surface. Platinum wires employed as suspenders, and coated along with the articles to be gilt, may also be cleaned without loss by making them anodes. If, on examination, all is found to be going on well, reimmerse the cathodes, and continue plating till they appear of a dull yellowish brown (this will occur in about four minutes), then remove them, rinse and scratch-brush them, and replace them in the bath.

When a second coat appears to be getting rather brown than yellowish brown, i.e. of the colour of wet wash-leather, the removal, followed by scratch-brushing, may be repeated, and for nearly all laboratory purposes, the articles are now fully gilt.

The coating of gold deposited from a hot cyanide solution is spongy in the extreme, and if the maximum wear-resisting effect is to be obtained, it is advisable to burnish the gold rather than to rely upon the scratch brush alone.

If the area of the cathode exceeds that of the anode the solution is said to grow weaker, and vice versa. This may be remedied in the former case by an obvious readjustment; the latter introduces no difficulty so far as I know except when plating iron or steel.

The student need not be troubled at the poor appearance of the deposit before it is scratch-brushed. Heavy gold deposits are almost always dull, not to say dirty, in appearance till the burnisher or scratch brush is applied. On the other hand, the deposit ought not to get anything like black in colour.

The following indications of defects may be noted--they are taken from Gore. I have never been really troubled with them.

The deposit is blackish. This is caused by too strong a current in too weak a bath. This may be remedied to some extent by stirring or keeping the cathode in motion. The obvious remedy is to add a little cyanide of gold.

The gold anode gets incrusted. This is a sign that the bath is deficient in potassium cyanide. The gold anode gets black and gives off gas. The solution is deficient in cyanide, and too large a current is being passed.

If a bright surface is desired direct from the bath, some caustic potash (say 2 per cent) may, according to Gore, be added, or the articles may be plated only slightly by using a weak current and taking them out directly they show signs of getting dull. By a weak current I mean one of about five amperes per square foot.

The deposit is said to be denser if the solution be heated as directed; but the bath will gild, though not quite so freely when cold.

To gild iron or steel directly, dilute the bath as above recommended some five or six times, add about 1 per cent of potassium cyanide, and gild with a very weak current (say two or three amperes per square foot) in the cold. Frequent scratch-brushing will be found requisite to secure proper adherence.

It is generally recommended to gild brass or German silver in solutions which are rather weak, but in the small practice which occurs in the laboratory a solution prepared as suggested does perfectly for everything except iron or steel. The scratch-brushing should be done over a large photographic developing dish to avoid loss of gold. It is a good plan to rinse the articles after leaving the bath in a limited quantity of distilled water, which is afterwards placed in a "residue" bottle, and then to scratch-brush them by hand over the dish to catch fine gold. When any loose dust is removed the articles may be scratched in the lathe without appreciable further loss.

Silver-gilt articles tend to get discoloured by use, but this discoloration can be removed by soap and water. After long use a gold cyanide bath tends to alter greatly in composition, In general, the bath tends to grow weaker, from the fact that there is a strong temptation to gild as many articles at once as possible.

It is therefore a good plan to keep a rough profit and loss account of the gold in order to find the quantity in solution. Fifty dwts. per gallon (or 78 grms. per 4.5 litres) is recommended. A gallon of solution of this strength is worth about eleven pounds sterling in gold and cyanide, and a serviceable anode will be worth about 10 pounds. (Fine gold is worth nominally four pounds four shillings and eleven pence ha'penny per oz.) Gold may be easily obtained containing less impurity than one part in ten thousand.

137. Plating with Copper. —

Copper may be deposited from almost any of its salts in reguline form, the sulphate and nitrate being most usually employed. In the laboratory a nearly saturated solution of sulphate of copper with 1 or 2 per cent of sulphuric acid will answer most purposes. A current density of, at most, fifteen amperes per square foot may be used, either for obtaining solid deposits for constructional purposes or for calibrating current measuring instruments by electrolysis. A copper anode is of course employed.

When coppering with a view to obtaining thick deposits it is a good plan to place the electrodes several inches apart, and, if possible, to keep the liquid stirred, as there is a considerable tendency on the part of copper deposits to grow out into mossy masses wherever the current density exceeds the limit mentioned. As the masses grow towards the anode the defect naturally tends to increase of itself, hence the necessity for care. The phenomenon is particularly marked at the edges and corners of the cathode.

If the deposit becomes markedly irregular, the best plan is to stop the process and file the face of the deposit down to approximate smoothness. In coppering it is of the utmost importance that the cathode be clean and free from grease; it must never be touched (by the finger, for instance) from the time it is scratch-brushed till it is immersed in the plating bath. Any grease or oxidation tends to prevent the copper deposit adhering properly.

A copper deposit oxidises very easily when exposed to the air. Consequently if the surface be required free from oxide, as, for instance, when it is to be silvered or gilt, it must be quickly washed when withdrawn from the coppering bath, scratch-brushed, and transferred immediately to the silvering or gilding bath.

If the surface is to be dried, as in electrolysis calibrations, it must be rinsed quickly with boiling water and pressed between sheets of filter paper. Another method which has been recommended is to rinse the copper in water slightly acidulated with sulphuric acid (which prevents oxidation), then in distilled water, and to dry by blotting paper and in front of a fire, taking care not to make the plate too hot. The wash water is sufficiently acidulated by the addition of two or three drops of acid per litre. So far as I know, the method of washing in acidulated water was first proposed by Mr. T. Gray.

138. Coppering Aluminium. —

A good adherent deposit of copper on aluminium used to be considered a desideratum in the days when it afforded the only means of soldering the latter. Many receipts have been published from time to time, and I have tried, I think, most of them. On no occasion, however, till this year (1896), have I succeeded in obtaining a deposit which would not strip after it was tinned and soldered, though it is not difficult to get apparently adherent deposits so long as they are not operated upon by the soldering iron. The best of the many solutions which have been proposed in years gone by is very dilute cupric nitrate with about 5 per cent of free nitric acid.

The problem of electroplating aluminium which I have indicated as awaiting a solution has at last found one. In the *Archives des Sciences physiques et naturelles de Genève* for December 1895 (vol. xxxiv.

p. 563) there is a paper by M. Margot on the subject, which discloses a perfectly successful method of plating aluminium with copper. The paper itself deals in an interesting way with the theory of the matter — however, the result is as follows.

(1) The aluminium articles are boiled for a few minutes in a strong solution of ordinary washing soda. The aluminium surface is thus corroded somewhat, and rendered favourable to the deposit of an adherent film of copper. After removal from the soda solution the aluminium is well washed and brushed in running water.

(2) The articles are dipped for thirty seconds or so in a hot 5 per cent solution of pure hydrochloric acid.

(3) After dipping in the hydrochloric acid, the work is instantly plunged into clean water for about one second, so as to remove nearly, but not quite, all of the aluminium chloride.

(4) The work is transferred to a cold dilute (say 5 per cent) solution of cupric sulphate slightly acidulated with sulphuric acid. The degree of acidulation does not appear to be very important, but about one-tenth per cent of strong acid does well.

If the preliminary processes have been properly carried out the aluminium will become coated with copper, and the process is accompanied by the disengagement of gas. It appears to be a rule that if gas is not given off, the film of copper deposited is non-adherent. The work must be left in the copper sulphate solution till it has received a uniform coating of copper.

(5) When this is the case the work is removed — well washed so as to get rid of the rest of the aluminium chloride, and then electroplated by the battery in the ordinary copper sulphate bath.

If the operation (4) does not appear to give a uniform coat, or if gas is not evolved from every part of the aluminium surface, I find that operations (2) and (3) may be repeated without danger, provided that the dip in the hydrochloric acid is shortened to two or three seconds.

The copper layer obtained by Margot's method is perfectly adherent — even when used as a base for ordinary solder — though in this case it can be stripped if sufficient force is applied.

Since the solder recommended by M. Margot for aluminium contains zinc, it does not run well when used to unite aluminium to copper, brass, iron, etc. In this case, therefore, I have found the most advantageous method of soldering to be by way of a preliminary copper-plating.

The success of M. Margot's method depends in my experience on obtaining just the proper amount of aluminium chloride in contact with the aluminium when the latter is immersed in the copper sulphate solution.

139. The process of copper-plating from sulphate or nitrate may, according to Mr. Swan (Journal of the Royal Institution, 1892, p. 630), be considerably accelerated by the addition of a trace of gelatine to the solution. As success appears to depend upon hitting the exact percentage amount of the gelatine, which must in any case be but a fraction of one per cent, and as Mr. Swan refrains from stating what the amount is, I am unable to give more precise instructions. A few experiments made on the subject failed, doubtless through the gelatine content not having been rightly adjusted. Mr. Swan claims to be able to get a hard deposit of copper with a current density of 1000 amperes per square foot, but seems to recommend about one-tenth of that amount for general use.

The solution employed is a mixture of nitrate of copper and ammonium chloride — proportions not stated. Electrolytic copper, as generally prepared, is very pure, but this is a mere accident depending on the impurities which, as a rule, have to be got rid of. Electrolysis seems to have no effect in purifying from arsenic, for instance.

Roughly speaking, about 11 grms. of copper are deposited per ampere hour from cupric salt solutions. When the current density is too high the anode suffers by oxidation, and this introduces a large and very variable resistance into the circuit.

140. Alkaline Coppering Solution —

Coppering Base Metals. — It is often desirable to coat lead, zinc, pewter, iron, etc., with a firm and uniform layer of copper preparatory to gilding or silvering. If copper or brass articles are soldered with soft solder it is found that the solder does not become silvered or gilt along with the rest of the material, but remains uncoated and of an ugly dark colour. This defect is got over by giving a preliminary coating of copper.

This is done in an alkaline solution, generally containing cyanogen and ammonia. The following method has succeeded remarkably well with me. The receipt was taken originally from Gore's *Electro-metallurgy*, p. 208. A solution is made of 50 grms. of potassium cyanide (ordinary commercial, say, 75 per cent) and 30 grms. of sodium bisulphite in I.5 litres of water. Thirty-five grammes of cupric acetate are dissolved in a litre of water, and 20 cubic centimetres of the strongest liquid ammonia are added. The precipitate formed must be more or less dissolved to a strong blue solution. The cyanide and bisulphite solution is then added with warming till the blue colour is destroyed. This usually requires the exact amount of cyanide and bisulphite mentioned, but I have not found it essential to entirely destroy the colour.

The solution contains cuprocyanide of sodium and ammonium (?), which is not very soluble, and this salt tends to be deposited in granular crystalline masses on standing. However, at a temperature of 50 C. the above receipt gives an excellent coppering liquid, which will coat zinc with a fine reguline deposit. Brass or copper partly smeared with solder will receive a deposit of copper on the latter as well as on the former, and, moreover, a deposit which appears to be perfectly uniform.

In using the bath the anode tends, as a rule, to become incrusted, and this rapidly increases the resistance of the cell, so that the current falls off quickly. The articles should be scratch-brushed and plated for about two minutes with a current density of about ten amp航s per square foot.

As soon as the deposit begins to look red the articles are to be removed and rebrushed, after which the process may be continued. About five minutes' plating will give a copper deposit quite thick enough after scratch-brushing to allow of a very even gilding or silvering.

Aluminium appears to be fairly coated, but, as usual, the copper strips after soldering. Iron receives an excellent and adherent coat.

I do not think that the formation of a crust upon the anode can be entirely prevented. According to Gore, its formation is due to the solution being too poor in copper, but I have added a solution of the acetate of copper and ammonium till the colour was bright blue without in any way reducing the incrustation. If the solutions become violently blue it is perhaps as well to add a little more cyanide and bisulphite, but I have not found such an addition necessary. The process is one of the easiest and most satisfactory in electro-metallurgy.

141. Nickel-plating.—

An examination of several American samples of nickel-plated goods has disclosed that the coating of nickel is, as a rule, exceedingly thin. This is what one would expect from laboratory repetition of the processes employed.

Commercial practice in the matter of the composition of nickelling solutions appears to vary a good deal. Thin coatings of nickel may be readily given in a solution of the double sulphate of nickel and ammonia, which does rather better if slightly alkaline. Deposits from this solution, however, become gray if of any thickness, and, moreover, are-apt to flake off the work. The following solution has given very good results with me. It is mentioned, together with others, in the *Electrical Review*, 7th June 1895.

The ingredients are:-

> Nickel sulphate 5 parts
>
> Ammonia sufficient to neutralise the nickel salt.
>
> Ammonium tartrate 3.75 parts
>
> Tannin 0.025 parts
>
> Water 100 parts

The nickel sulphate and ammonia are dissolved in half the water, the ammonium tartrate in the other half with the tannin. The solutions are mixed and filtered at about 40 C. This solution works well at ordinary temperatures, or slightly warm, with a current density of ten amp航s per square foot. In an experiment made for the purpose I found that plating may go on for an hour in this solution before the deposit begins to show signs of flaking off. The deposit is of a fine white colour.

The resistance of the bath is rather high and rather variable, consequently it is as well to have a current indicator in circuit, and it may well happen that five or six volts will be found requisite to get the current up to the value stated. For nickelling small objects of brass, such as binding screws, etc., it is very necessary to be careful as to the state of polish and uniformity of their surfaces before placing them in the plating bath. A polished surface will appear when coated as a polished surface, and a mat surface as a mat surface; moreover, any local irregularity, such as a speck of a foreign metal, will give rise to an ugly spot in the nickelling bath. For this reason it is often advisable to commence with a coat of copper laid on in an alkaline solution and scratch-brushed to absolute uniformity.

An examination of the work will, however, disclose whether such a course is desirable or not; it is not done in American practice, at all events for small brass objects. These are cleaned in alkali and in boiling cyanide, which does not render a polished surface mat, as weak acid is apt to do, and are then coated with a current density of about ten amp航s per square foot.

In spite of what is to be found in books as to the ease with which nickel deposits may be polished, I find that the mat surface obtained by plating on an imperfectly polished cathode of iron is by no means easily polished either by fine emery, tripoli, or rouge. Consequently, as in the case of brass, if a polished surface is desired, it must be first prepared on the unplated cathode. In this case, even if the deposit appears dull, but not gray, it may be easily polished by tripoli and water, using a cork as the polisher. Scratch-brushing with brass wire, however, though possibly not with German silver wire, brightens the deposit, but discolours it. When the deposit becomes gray I have not succeeded in polishing it satisfactorily.

Soldered brass or iron may be satisfactorily coated with nickel by giving it a preliminary coating of copper in the cyanide bath. On the whole, I recommend in general that iron be first coated with copper in the alkaline bath, scratch-brushed, and then nickel-plated, and this whether the iron appears to be uniform or not. Much smoother, thicker, and stronger coats of nickel are obtained upon the copper-plated surface than on the iron one, and the coating does not become discoloured (? by iron rust) in the same way that a coating on bare iron does. The copper surface may be plated for at least an hour at a density of ten amp航s per square foot without scaling.

Scales or circles divided on brass may be greatly improved in durability by nickel-plating. For this purpose the brass must be highly polished and divided before it is nickelled.

The plating should be continued for a few minutes only, when a very bright but thin coat of nickel will be deposited; it then only remains to wash and dry the work, and this must be done at once. If the nickel is deposited before the scale or circle is engraved, very fine and legible divisions are obtained, but there is a risk that flakes of nickel may become detached here and there in the process of engraving.

142. Miscellaneous Notes on Electroplating.

Occasionally it is desirable to make a metallic mould or other object of complex shape. The quickest way to do this is to carve the object out of hard paraffin, and then copy it by electrotyping. Electrotype moulds can be made in many ways. The easiest way perhaps is to take a casting in plaster of Paris, or by means of pressure in warm gutta-percha.

In cases where the mould will not draw, recourse must be had to the devices of iron-founders, i.e. the plaster cast must be made in suitable pieces, and these afterwards fitted together. This process can occasionally be replaced by another in which the moulding material is a mixture of treacle and glue. The glue is soaked in cold water till it is completely soft. The superfluous water thrown away, one-fourth part by volume of thick treacle is added, and the mixture is melted on the water bath; during which process stirring has to be resorted to, to produce a uniform mixture.

This liquid forms the moulding mixture, and it is allowed to flow round the object to be copied, contained in a suitable box, whose sides have been slightly oiled. The object to be copied should also be oiled. After some hours, when the glue mixture has set, it will be found to be highly elastic, so that it may be pulled away from the mould, and afterwards resume very nearly its original form.

One drawback to the use of these moulds lies in the fact that the gelatine will rarely stand the plating solution without undergoing change, but this may be partially obviated by dipping it for a few seconds in a 10 per cent solution of bichromate of potash, exposing it to the sunlight for a few minutes, and then rinsing it.

In order to render the surface conducting, it is washed over with a solution of a gold or silver salt, and the latter reduced in situ to metal by a suitable reagent. A solution of phosphorus is the most usual one (see Gore, Electro-metallurgy, p. 216). Such a mould may be copper-plated in the sulphate bath, connection being made by wires suitably thrust into the material.

Plaster of Paris moulds require to be dried and waxed by standing on a hot plate in melted wax before they are immersed in the plating bath. In this case the surface is best made conducting either by silvering it by the silvering process used for mirrors, or by brushing it over with good black lead rendered more conducting by moistening with an ethereal solution of chloride of gold and then drying in the sun.

The brushing requires a stiff camel's-hair pencil of large size cut so that the hairs project to a distance of about a quarter of an inch from the holder. The brushing must continue till the surface is bright, and is often a lengthy process.

The same process of blackleading may be employed to get a coat of deposited metal which will strip easily from the cathode.

In all cases where extensive deposits of copper are required, the growth takes place too rapidly at the corners. Consequently it is often desirable to localise the action of the deposit. A "stopping" of ordinary copal varnish seems to be the usual thing, but a thin coat of wax or paraffin or photographic (black) varnish does practically as well.

I do not propose to deal with the subject of electrotyping to any extent, for if practised as an art, a good many little precautions are required, as the student may read in Gore's Electro-metallurgy. The above instructions will be found sufficient for the occasional use of the process in the construction of apparatus, etc. There is no advantage in attempting to hurry the process, a current density of about ten amp航s per square foot being quite suitable and sufficiently low to ensure a solid deposit.

143. Blacking Brass Surfaces. —

A really uniform dead-black surface is difficult to produce on brass by chemical means. A paste of nitrate of copper and nitrate of silver heated on the brass is said to give a dead-black surface, but I have not succeeded in making it act uniformly. For optical purposes the best plan is to use a paint made up of "drop" black, ground very fine with a little shellac varnish, and diluted for use with alcohol. No more varnish than is necessary to cause the black to hold together should be employed.

In general, if the paint be ground to the consistency of very thick cream with ordinary shellac varnish it will be found to work well when reduced by alcohol to a free painting consistency.

A very fine gray and black finish, with a rather metallic lustre, may be easily given to brass work. For this purpose a dilute solution of platinum tetrachloride (not stronger than 1 per cent) is prepared by dissolving the salt in distilled water. The polished brass work is cleaned by rubbing with a cork and strong potash till all grease has disappeared, as shown by water standing uniformly on the metal and draining away without gathering into drops.

After copious washing the work is wholly immersed in a considerable volume of the platinum tetrachloride solution at the ordinary temperature. After about a quarter of an hour the brass may be taken out and washed. The surface will be found to be nicely and uniformly coated if the above instructions have been carried out, but any finger-marks or otherwise dirty places will cause irregularity of deposit. If the process has been successful it will be found that the deposit adheres perfectly, hardly any of it being removed by vigorous rubbing with a cloth. If the deposit is allowed to thicken — either by leaving the articles in the solution too long or heating the solution, or having it too strong — it will merely rub off and leave an irregular surface.

This process succeeds well with yellow brass and Muntz metal, either cast or rolled, but it does not give quite such uniform (though still good) results with gun-metal, on which, however, the deposit is darker and deader in appearance.

A book might be written (several have been written) on the art of metal colouring, but though doubtless a beautiful and delicate art, it is of little service in the laboratory. For further information the reader may consult a work by Hiorns.

144. *Sieves.* —

Properly graded sieves with meshes of a reliable size are often of great use. They should be made out of proper "bolting" cloth, a beautiful material made for flour-millers. Messrs. Henry Simon and Company of Manchester have kindly furnished me with the following table of materials used in flour-milling.

Sieves made of these materials will be found to work much more quickly and satisfactorily than those made from ordinary muslin or wire gauze.

Relative Bolting Value of Silk, Wire, and Grit Gauze

Threads per inch Approximate.	Trade No. of Silk.	Trade No. of Wire.	Trade No. of Grit Gauze.
18	0000	18	16
22	000	20	20
28	00	26	26
38	0	32	34

48	1	40	44
52	2	45	50
56	3	50	54
60	4	56	58
64	5	60	60
72	6	64	66
80	7	70	70
84	8	80	80
94	9		
106	10		
114	11		
124	12		
130	13		
139	14		

148	15
156	16
163	17
167	18
170	19
173	20

145. Pottery making in the Laboratory.

—

When large pieces of earthenware of any special design are required, recourse must be had to a pottery. Small vessels, plates, parts of machines, etc., can often be made in the laboratory in less time than it would take to explain to the potter what is required. For this purpose any good pipeclay may be employed. I have used a white pipe-clay dug up in the laboratory garden with complete success.

The clay should be kneaded with water and squeezed through a cloth to separate grit. It is then mixed with its own volume or thereabouts of powdered porcelain evaporating basins, broken basins being kept for this purpose. The smoothness of the resulting earthenware will depend on the fineness to which the porcelain fragments have been reduced. I have found that fragments passing a sieve of sixty threads to the inch run, do very well, though the resulting earthenware is decidedly rough.

The porcelain and clay being thoroughly incorporated by kneading, the articles are moulded, it being borne in mind that they will contract somewhat on firing. *[Footnote:* The contraction depends on the temperature attained as well as on the time. An allowance of one part in twelve will be suitable in the case considered.] The clay should be as stiff as is convenient to work, and after moulding must be allowed to get thoroughly dry by standing in an airy place; the drying must not be forced, especially at first, or the clay will crack.

Small articles are readily fired in a Fletcher's crucible furnace supplied with a gas blow-pipe; the furnace is heated gradually to begin with. When a dull red heat is attained, the full power of the blast may be turned on, and the furnace kept at its maximum temperature for three or four hours at least, though on an emergency shorter periods may be made to do.

The articles are supported on a bed of white sand; after firing, the crucible furnace must be allowed to cool slowly. It must be remembered that the furnace walls will get hot externally after the first few hours, consequently the furnace must be supported on bricks, to protect the bench.

The pottery when cold may be dressed on a grindstone if necessary. This amateur pottery will be found of service in making small fittings for switch-boards, commutators, and in electrical work generally.

Pottery made as described is very hard and strong, the hardness and strength depending in a great degree on the proportion of powdered porcelain added to the clay, as well, of course, as on the quality of both of these materials.

It is a good plan to knead a considerable quantity of the mixture, which may then be placed in a well-covered jar, and kept damp by the addition of a little water.

Pottery thus made does not require to be glazed, but, of course, a glaze can be obtained by any of the methods described in works on pottery manufacture. The following glaze has been recommended to me by a very competent potter:—

Litharge   7 parts by weight

Ground flint   2 parts by weight

Cornish stone or   1 parts by weight
felspar

These ingredients are to be ground up till they will pass the finest sieve — say 180 threads to the inch. They are then mixed with water till they form a paste of the consistency of cream. They must, of course, be mixed together perfectly. The ware to be glazed is dipped into the cream after the first firing; it is then dried as before and refired. The glaze will melt at a bright red heat, but it will crack if not fired harder; the harder it is fired the less likely is it to crack.

If colouring matters are added they must be ground in a mill free from iron till they are so fine that a thick blanket filter will not filter them when suspended in water. This remark applies particularly to oxide of cobalt.

APPENDIX

IN the *Philosophical Magazine* for July 1888 (vol. xxvi. p. 1) there is a paper by Professor Kundt translated from the *Sitzungsberichte* of the Prussian Academy. This paper deals with the indices of refraction of metals. Thin prisms were obtained by depositing metals electrolytically on glass surfaces coated with platinum. The preparation of these surfaces is troublesome. Kundt recounts that no less than two thousand trials were made before success was attained. A detailed account of the preparation of these surfaces is not given by Kundt, but one is promised — a promise unfortunately unfulfilled so far as I am able to discover. A hunt through the literature led to the discovery of the following references: *Central Zeitung fuer Optik und Mechanik*, p. 142 (1888); *Dingler's Polytechnik Journal*, Vol. cxcv. p. 464; *Comptes Rendus*, vol. lxx. (1870).

The original communication is a paper by Jouglet in the Comptes Rendus, of which the other references are abstracts. The account in Dingier is a literal translation of the original paper, and the note in the *Central Zeitung* is abbreviated sufficiently to be of no value. The details are briefly as follows:-

One hundred grams of platinum are dissolved in aqua regia and the solution is dried on the sand bath, without, however, producing decomposition. Though the instructions are not definite, I presume that the formation of $PtCl_4$ is contemplated.

The dried salt is added little by little to rectified oil of lavender, placed on a glass paint-grinding plate, and the salt and oil are ground together with a muller. Care is required to prevent any appreciable rise of temperature which would decompose the compound aimed at, and it is for this reason that the salt is to be added gradually. Of course the absorption of water from the air must be prevented from taking place as far as possible. Finally, the compound is diluted by adding oil of lavender up to a total weight of 1400 grams (of oil).

The liquid is poured into a porcelain dish and left absolutely at rest for eight days. It is then decanted and filtered, left six days at rest, and again decanted (if necessary). The liquid should have a specific gravity of 5 on the acid hydrometer. (If by this the Baum頼cale is intended, the corresponding specific gravity would be 1.037.) A second liquid is prepared by grinding up 25 grams of litharge with 25 grams of borate of lead and 8 to 10 grams of oil of lavender. The grinding must be thoroughly carried out.

This liquid is to be added to the one first described, and the whole well mixed. The resulting fluid constitutes the platinising liquid, and is applied as follows:-

A sheet of clean glass is held vertically, and the liquid is painted over it, carrying the brush from the lower to the upper edge. The layer of oil dries slowly, and when it is dry the painting is again proceeded with, moving the brush this time from right to left; and similarly the process is repeated twice, the brush being carried from top to bottom and left to right. This is with the object of securing great uniformity in the coating. Nothing is said as to the manner in which the glass is to be dried.

The dried glass is finally heated to a temperature of dull redness in a muffle furnace. The resinous layer burns away without running or bubbling, and leaves a dull metallic surface. As the temperature rises this suddenly brightens, and we obtain the desired surface (which probably consists of an alloy of lead and platinum). It is bright only on the surface away from the glass.

I have not had an opportunity of trying this process since I discovered the detailed account given by Jouglet; but many modifications have been tried in the laboratory of the Sydney University by Mr. Pollock, starting from the imperfect note in the Central Zeitung, which led to no real success.

It was found that it is perfectly easy to obtain brilliant films of platinum by the following process, provided that the presence of a few pin-holes does not matter.

The platinum salt employed is what is bought under the name of platinic chloride; it is, however, probably a mixture of this salt and hydro-chloro-platinic acid, and has all the appearance of having been obtained by evaporating a solution of platinum in aqua regia to dryness on the water bath. A solution of this salt in distilled water is prepared; the strength does not seem to matter very much, but perhaps one of salt to ninety-nine water may be regarded as a standard proportion. To this solution is added a few drops of ordinary gum water (i.e. a solution of dextrin). The exact quantity does not matter, but perhaps about 2 per cent may be mentioned as giving good results.

The glass is painted over with this solution and dried slowly on the water bath. When the glass is dry, and covered with a uniform hard film of gum and platinum salt free from bubble holes, it is heated to redness in a muffle furnace. The necessary and sufficient temperature is reached as soon as the glass is just sensibly red-hot.

The mirrors obtained in this way are very brilliant on the free platinum surface. If the gum be omitted, the platinum will have a mat surface; and if too much gum be used, the platinum will get spotty by bubbles bursting. There does not appear to be any advantage in using lead.

It is quite essential that the film be dry and hard before the glass is fired.